The Satellite Technology Guide for the 21ˢᵗ Century

To David —

Thanks for all great times, memories and support over the years.

You and the GVF have definitely made a positive impact in the VSAT industry.

All the very best,

John Puetz

2/25/08

Also by Virgil S. Labrador:

with Peter I. Galace. *Heavens Filled with Commerce: A Brief History of the Communications Satellite Industry.* Satnews Publishers, 2005.

with Florangel Braid, Ramon Tuazon, Elizabeth Dimasuay and Ma. Imelda Samson. *Communication for the Common Good: Towards a Framework for a National Communication Policy.* Asian Institute of Journalism, 1990.

with Florangel Braid, Evangeline R. Alberto, Ramon Tuazon, Elizabeth Dimasuay and Ma. Imelda Samson. *A Resource Guide and Module on Corporate Communication.* Asian Institute of Journalism, 1990.

with Angela Mia Serra. *Basic Course in Community Communications Training Manual.* PROCESS, 1987.

The Satellite Technology Guide for the 21st Century

by Virgil S. Labrador

with chapter contributions from
**John M. Puetz, DC Palter
and Daniel B. Freyer**

SYNTHESIS PUBLICATIONS
Los Angeles, California USA

Published by:

SYNTHESIS PUBLICATIONS
Los Angeles, California, USA

Cover Design, Layout and Illustrations
by Peter I. Galace

Copyright © 2008

All rights reserved. Reproduction or publication of the contents of this book in any manner without prior written permission of the publisher is prohibited except in the case of brief quotations embodied in critical articles and reviews. Address all correspondence to: Synthesis Publications, P.O. Box 4174, West Covina, CA 91791-0174 USA.

ISBN No. 978-1-60530-421-2

Printed in the United States of America.

Preface

The Satellite Technology Guide for the 21st Century is a brief overview of the global satellite communications industry and how satellite communications technology works. The book is aimed at a non-technical audience who would like to explore the subject, or fill in some gaps in their understanding, of satellite communications technology and the industry. The idea is to explain the technology in easy to understand terms and relate it to the current structure and composition of the global satellite industry.

While the basic principles of satellite technology remain constant, the satellite industry is one of the most dynamic sectors of the global economy. The industry has been experiencing major changes and unprecedented growth in the last few years since the advent of the new millennium. This book aims to provide not just a technical primer but put technology in the context of how it is currently applied and developed by the industry. To give as broad a view of the subject as possible, I have brought in leading industry experts such as Daniel Freyer, John Puetz and DC Palter to write chapters about key industry sectors such as Satellite Services, VSATs and Satellites and the Internet, respectively.

This book is organized into nine chapters that individually focus on a key topic. The chapters are arranged in a logical sequence but can also be read in any order or by itself as each chapter delves on a specific aspect of the technology and the industry. At the back of the book is an Appendix section that includes a handy glossary of terms that explains industry jargon and the major terms used in this book. There is also a listing of important industry resources where you can get more information and suggested further reading.

Acknowledgements

I would like to express my deepest gratitude and appreciation to my aforementioned co-authors, Daniel Freyer, John Puetz and DC Palter for their excellent chapter contributions to this book. Without their expertise and experience this book would be a very different product. Thanks to Peter Galace, who was my co-author of the book on the history of the satellite industry and long-time colleague and friend. Peter did all the layout and graphic work for this book. His graphics helped demystify complex technology and concepts and the layout provided the unique overall 'look' of the book.

There are also numerous individuals in the industry that I have worked with over the years, from whom I have benefited from their knowledge and experience, notably Bruce Elbert of Application Technology Strategy, Thomas van der Heyden and Howard Greenfield. Their wise counsel and support contributed significantly to my knowledge and appreciation of this great industry.

Finally, I can't express enough my thanks to my family: my wife Jackie, my son Carlo, my mother Josephine and my brother John Paul for putting up with my omissions during the arduous process of completing this book and for providing the inspiration and motivation to bring it to fruition.

Virgil Labrador

January 2008
Los Angeles, California

CONTENTS

Chapter 1: Brief History of the Satellite Communications Industry — 11
by Virgil S. Labrador

Arthur C. Clarke's Idea • Syncom—first geostationary satellite • Birth of Intelsat • Competition and Consolidation

Chapter 2: Overview of Communications Satellite Industry — 21
by Virgil S. Labrador

Main Segments of the Satellite Industry • Satellite Manufacturing • Satellite Launch Providers • Satellite Service Providers • Ground Equipment Manufacturers • Jobs in the Industry

Chapter 3: The Basics—*by Virgil S. Labrador* — 37

A Model of the Communication Process • Electromagnetic Spectrum • Orbital Locations • Satellite Frequency Bands • Basic Satellite Network Architecture

Chapter 4: The Space Segment—*by Virgil S. Labrador* — 55

Types of Satellites • Spin Stabilized vs. Body Stabilized Satellites • Satellite Applications • Satellite Bus • Communications Payload • Launching a Satellite • Satellite Insurance

Chapter 5: The Ground Segment—*by Virgil S. Labrador* — 77

Earth Stations • Earth Station Architecture • Teleports • Teleport Services • Satellite Networks • Advantages and Disadvantages of Satellite Communications

Chapter 6: Satellite Services—*by Daniel B. Freyer* — 93

Analog vs. Digital • Fixed Satellite Services • Broadcast Services • Telecommunications Services • Direct-to-Home Services • Broadband Services • New and Emerging Services

Chapter 7: VSATs—*by John M. Puetz* — 119

Why VSATs? • Brief History of VSATs • System Topology and Satellite Access Methods • Networking, Routing and Security • Fixed vs. Mobile VSATs • VSAT Applications • Emerging Trends

Chapter 8: Satellites and the Internet—*by DC Palter* — 157

IP Networks • Hybrid Networks • Internet Protocols • Broadband Services and Standards •

Chapter 9: The Future of Satellite Communications— *by Virgil S. Labrador* — 175

Consolidation • New and Emerging Technologies and Applications • Satellite Markets • Industry Trends • Competing Technologies • What's in store?

APPENDICES — 181

Glossary of Terms

Recommended Further Reading

Industry Resources

"Any sufficiently advanced technology is indistinguishable from magic."

-Arthur C. Clarke

x

CHAPTER 1

A Brief History of the Commercial Satellite Industry

The commercial satellite industry is just a little over forty years old, having started operations with the founding of the first satellite company, Intelsat, in 1964. Although sages and novelists have been dreaming and writing of satellites and space travel for many centuries, the workable idea for a man-made satellite did not come about until near the end of World War II. Satellite technology was conceived by a 27-year old Royal Air Force (RAF) officer named Arthur C. Clarke in 1945 in his spare time while serving as a training officer in the remote English countryside.

Clarke wrote a letter that provided the blueprint for communication satellites that was published in the February 1945 issue of the British magazine, *Wireless World*. His letter goes into details about the concept of a satellite, to wit:

"...An 'artificial satellite' at the correct distance from the earth will make a revolution every 24 hours, i.e. it would remain stationary about the same spot and would be within optical range of nearly half the earth's surface...Three repeater stations, 120 degrees apart in the correct orbit, could give television and microwave coverage to the entire planet..."

The distance from the earth where a satellite will be moving at the same speed as the earth's rotation was determined by Clarke to be 22,300 miles (35,888 kms.) above the earth's surface. By moving at the same speed as the earth, the satellite will always be on the same fixed position above relative to a position on the ground. This would mean, that a receiver or transmitter on earth can be pointed directly to the satellite 24 hours a day, precluding the need for multiple ground stations to track the satellite. This orbital arc is now officially called "Clarke's Orbit" by the International Astronomical Union.

Figure 1.1: Satellite Earth Orbits

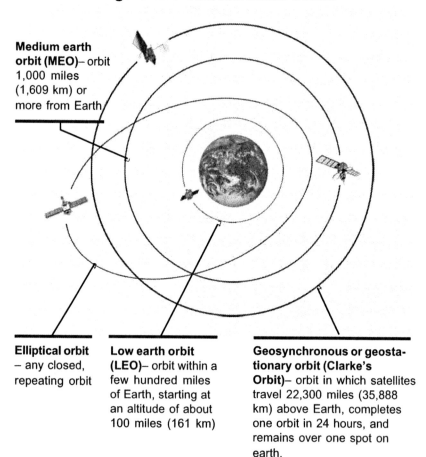

Medium earth orbit (MEO)– orbit 1,000 miles (1,609 km) or more from Earth

Elliptical orbit – any closed, repeating orbit

Low earth orbit (LEO)– orbit within a few hundred miles of Earth, starting at an altitude of about 100 miles (161 km)

Geosynchronous or geostationary orbit (Clarke's Orbit)– orbit in which satellites travel 22,300 miles (35,888 km) above Earth, completes one orbit in 24 hours, and remains over one spot on earth.

In the same letter, Clarke mentioned that "an 'artificial satellite' would be a possibility in the more remote future—perhaps—*half a century ahead* "(emphasis mine). By his initial reckoning, it won't be till 1995 before a working satellite would have been actually developed and launched. Clarke was thinking that the satellite would be enormous in order to accommodate all the vacuum-tube electronics needed and will have to be manned. While he made the quantum leap by providing the blueprint that would make satellite technology feasible, he did not foresee the development of the transistor and subsequent advancements in miniaturization that made possible the launch of a satellite just 12 years after his landmark article, with the successfully placing into orbit of the Russian "Sputnik" satellite in 1957.

Clarke later wrote a memo on May 25, 1945 to a fellow member of the British Interplanetary Society, Ralph Slazenger of the Slazenger sporting goods company. The four-page memo entitled "The Space Station: Its Radio Applications," outlined the feasibility of an orbiting space station and its use as a communications medium (the original copy of this memo is now permanently displayed in the Smithsonian Museum in Washington, D.C.). He then wrote a more detailed paper published in the October 1945 issue of *Wireless World* entitled "Extra-Terrestrial Relays: Can Rocket Stations Give World-wide Radio Coverage?" When Clarke submitted the article, he originally gave it an ominous title: "The Future of World Communications," but the editors decided on the aforementioned more modest title.

Arthur C. Clarke as a young RAF officer in 1943.

In hindsight, Clarke's original title might have been more appropriate. The article is widely considered a landmark work that was to launch the commercial satellite industry as we know it today and changed forever how the world communicates and relates to each other.

After World War II, the resulting cold war between the U.S. and the Soviet Union spurred an arms race that, while wasteful and permeated an ominous pall over the world, resulted in many technological innovations that would eventually have useful civilian applications. One of them was the development of more powerful long-range rockets. In its desire to be able to launch a rocket with warheads that can reach the continental United States, the Soviet Union embarked on an ambitious development program for Intercontinental Ballistic Missiles (ICBMs). The same rockets, however, can also be used to launch artificial satellites into space.

In October 1954, an initiative led by the science academies of 67 nations including the U.S. and the USSR adopted a resolution to launch an artificial satellite to map the earth's surface during the International Geophysical Year (IGY) that began on July 1, 1957. The IGY actually lasted 18 months from July 1, 1957 to December 31, 1958. Both the US and the USSR announced their intention to launch an artificial satellite during the IGY. The U.S. actually was able to launch a dummy payload filled with sand on a Jupiter C rocket on September 20, 1956. The rocket was able to reach an altitude of 682 miles but stopped short of boosting its final stage into orbit. The explanation for the dummy payload was ostensibly to preserve the scientific nature (as opposed to military) of the project in accordance with the IGY's objectives. This would have beaten the Soviet Union by over a year, which managed to launch a beach-ball sized artificial satellite called Sputnik on October 4, 1957.

The successful launch of Sputnik shocked the Western world and established Soviet technological leadership at the time. It was to result in a space race that culminated in the U.S. landing the first humans on the moon on July 20, 1969. After several failed attempts, the US finally launched its first satellite—Explorer 1 on January 31, 1958. The satellite carried a small scientific payload that eventually discovered the magnetic radiation belts around the earth—the Van Allen Belt, named after its principal investigator, astrophysicist James Van Allen.

Directly resulting from the backlash of the Sputnik launch, the US government created on February 7, 1958 the Advanced Research Projects Agency (ARPA). The agency was mandated to ensure US technological leadership and prevent future "Sputniks." ARPA was to make many technological breakthroughs including the ARPANET computer network in the '60s—a precursor to the internet.

The first project to be undertaken by the newly-created ARPA was Project SCORE (Signal Communication by Orbiting Relay Equipment) in cooperation with the US Army Signal Corps and the Air Force. SCORE launched the first communications satellite from Cape Canaveral, Florida on December 19, 1958. The satellite broadcasted a taped message from US President Dwight D. Eisenhower.

The practical development of communications satellite technology is largely credited to two engineers, Dr. John Pierce of Bell Labs and Dr. Harold Rosen of Hughes Aircraft Company (later know as Hughes Space and Communications and now Boeing Satellite Systems). Dr. Pierce's main contribution was the development of the Traveling Wave Tube (TWT) used in satellites, which is capable of receiving and transmitting radio signals (the precursor of today's transponders). While Dr. Rosen developed spin-stabilization technology that provides stability to satellites orbiting in space.

Dr. Pierce led a team at Bell Labs that launched the first passive communications satellite Echo-1 on August 12, 1960. Echo-1 was a 100-foot sphere which contained no instruments on its own but was able to reflect signals from the ground, thus it was called a "passive" satellite. Dr. Pierce's team at Bell Labs also launched the first active communications satellite into low earth orbit, Telstar 1 on July 10, 1962. Telstar 1 was the first satellite to transmit live television images from Europe to North America and vice-versa. Telstar 1 also transmitted the first phone call via satellite—a brief call from AT&T Chairman Fred Koppel from its ground station in Andover, Maine to US President Lyndon Johnson in Washington, D.C.

Dr. Rosen's team at Hughes Aircraft launched the first geosynchronous satellite, Syncom-1 on February 14, 1963. However, Syncom-1 was lost shortly after it was successfully placed in geostationary orbit. Geosynchronous satellites, as Clarke conceived it in 1945, move at the same speed as the earth's orbit, making it appear stationary from a position in the ground. This precluded the need for multiple tracking stations and also increased the coverage of the satellite to up to one-third of the earth's surface.

Syncom-1 was followed by the successful launch of Syncom-2 in July 26, 1963. However, Syncom-2 was placed in an inclined orbit, 33.05 degrees from the equator. The first fully-operational communications satellite successfully launched into Clarke's geostationary orbit was Syncom-3 in August 19, 1964. Syncom-3 was able to broadcast live the Tokyo Olympics from Japan to the US.

The successful development of a geostationary satellite and its potential use for telecommunications and broadcasting paved the way for a global communications satellite industry. The US spearheaded the development of such an industry with the passing of the Communications Satellite Act in 1962 which authorized the formation of the Communications Satellite Corporation (Comsat), a private company that will represent the US in a new international satellite communications organization to be called "Intelsat."

Intelsat was formed on August 20, 1964 with 11 signatories to the Intelsat Interim Agreement. The original 11 signatories included the US (represented by Comsat), United Kingdom (represented by the Government Post Office), Canada, Japan, the Netherlands, Germany, Spain,

The Satellite Technology Guide for the 21st Century

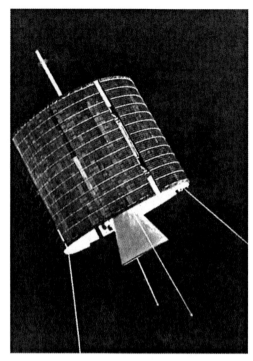

The first commercially operating satellite in geostationary orbit, Early Bird, launched on April 6, 1965. Today all satellites are generically referred to as "birds." (Photo courtesy of INTELSAT)

Austria, Norway, Switzerland (represented respectively by their Post and Telegraph authorities or PTTs) and the Vatican City. Intelsat launched on April 6, 1965 its first satellite, Early Bird—designed and built by Dr. Rosen's team at Hughes Aircraft Company. Early Bird was the first operational commercial satellite providing regular telecommunications and broadcasting services from North America to Europe and vice-versa. To this day, communications satellites are generically referred to as "birds" after the pioneering achievement of the Early Bird satellite.

Early Bird was followed in 1967 by Intelsat 2 for the Pacific Ocean region and on July 1, 1969, Intelsat III covering the Indian Ocean region was launched, thereby achieving full global coverage. 19 days later, the landing of the first human on the moon was broadcasted live through the global network of three Intelsat satellites to over 500 million television viewers on earth.

The potential uses for development and reaching out to remote regions led other countries to build and operate their own national satellite systems. Canada was the first country after the US and the USSR to

A Brief History of the Commercial Satellite Industry

launch its own satellite, Anik 1, on November 9, 1972. This was followed by Indonesia's Palapa satellite on July 8, 1976. Many other countries followed suit and launched their own satellites.

To counteract Intelsat, the then Eastern Bloc countries under Soviet control formed their own international satellite consortium called "Intersputnik" in 1974. Until the early 80s, Intelsat and Intersputnik had a virtual monopoly on international satellite communications.

In 1984 a colorful entrepreneur named Rene Anselmo sought out to challenge Intelsat's monopoly of international satellite communications and formed a company called Alpha Lyracom/Pan American Satellite, later shortened to PanAmSat. Using unorthodox tactics such as taking out full-page ads featuring an irreverent dog named "Spot" in the *Wall Street Journal* and satellite trade publications, Anselmo lobbied the US Federal Communications Commission (FCC) to open the satellite communications market to competition. Anselmo's efforts met with success and on June 15, 1988, PanAmSat launched its first satellite dubbed "PAS-1."

PanAmSat paved the way for many private operators to flourish in the international satellite market such as Luxembourg's Sociéte Européenne de Satelites (SES) which launched Astra 1A on December 11, 1988. Astra 1A carried four channels of Rupert Murdoch's fledging Direct-to-Home (DTH) satellite service called Sky. SES was a pioneer in the DTH business in Europe and the company grew through the 90s, challenging market leaders Intelsat and PanAmSat. Meanwhile, a Hong Kong-based consortium launched AsiaSat 1 satellite in April 7, 1990 introducing DTH services in the Asian market.

By the mid 90s, it was evident that Intelsat's leading position in the international communications satellite market was eroding fast with the rise of privately-owned operators PanAmSat and SES and the entry of many new players such as GE Americom, Orion Network Systems, Columbia Communications Systems and many other domestic operators in Europe, Asia and Latin America. The industry was by now a global industry and increasingly dominated by private corporations that were not faced by the same constraints as Intelsat, which has grown to an unwieldy international bureaucracy with 144 country member-signatories. In 1996, Intelsat decided that privatization was the only way to go and began the process to transform the organization from an essentially inter-governmental body to a private corporation.

Meanwhile in 1998, Intelsat spun off a new private entity called

The Satellite Technology Guide for the 21st Century

Representatives from 11 countries that were the original signatories to the agreement creating Intelsat signed on August 20, 1964. The formation of Intelsat marked the beginning of the commercial communications satellite industry as we know it today. (photo courtesy of Intelsat)

New Skies Satellites based in The Hague, Netherlands, while its own privatization process was underway. Intelsat officially became a private corporation on July 18, 2001, four months after SES, now called SES Global, surpassed Intelsat as the largest satellite operator in world, after it acquired US operator, GE Americom on March 28, 2001.

With Intelsat fully privatized, the competition in the international satellite market heated up. After the telecom and dot.com bubble burst in the late 90s, the industry underwent a period of consolidation and mergers which culminated in August 2005, when Intelsat announced that it is acquiring PanAmSat for $3.2 Billion. The combined Intelsat-PanAmSat company will have 53 satellites worldwide and close to $2 Billion in annual revenues, making it once again the largest satellite operating company in the world. Not to be undone, SES Global announced in December 2005 that it is acquiring the erstwhile Intelsat spin-off, News Skies Satellite, thus narrowing the gap with leading operator Intelsat.

The process of consolidation continues while the global satellite industry continues to grow. According to the Washington, D.C.-based Satellite Industry Association (SIA) which does an annual survey of the industry, the industry has been growing consistently even during the dot.com and telecom busts of the late 90s at an average rate of about 10 percent. In 2006, the global satellite industry surpassed US $100 Billion in revenues.

It has certainly come a long way since a young budding science-fiction writer named Arthur C. Clarke first conceived it in 1945.[1]

---

[1] For a more detailed account of the history of the commercial satellite communications industry, see the book "*Heavens Filled with Commerce: A Brief History of the Satellite Communications Industry*" by Virgil S. Labrador and Peter I. Galace, published by Satnews Publishers, 2005 (for more info go to www.satnews.com/products/historybook.htm).

CHAPTER 2

Overview of the Satellite Communications Industry

The satellite communications industry in 2006 generated revenues of US$ 106.1 Billion globally according to the Washington, D.C.- based Satellite Industry Association (SIA). Thus breaking the US $100 Billion mark for the first time in its relatively short history. The worldwide satellite industry has been growing an average of 10 per cent per annum ever since the SIA and research company Futron Corporation began tracking the industry in 1996. This growth was sustained even during the economic downturn of the late '90s and early 2000s due to the internet and telecom busts. In June 2007, at the ISCe Conference in San Diego, California, SIA Executive Director David Cavossa said that "this growth is expected to continue, driven by growing demand for continuity of communications and the renewed interest by financial markets to invest in satellite companies."

The continued growth of the satellite industry despite the many challenges and the turmoil that plague the telecommunications industry as whole in the last few years makes it a very good industry to be in at this point in time. Given its relatively young history, having just started in 1964 when Intelsat was formed, the industry is yet to reach full maturity and thus still has a lot of room for growth. The experience in the late 90s when the dot.com bubble burst and the telecommunications industry was plagued with high profile busts and scandals, only serve to strengthen the industry as it takes a more "back to the basics" approach in the new millennium. The satellite industry is sticking to what made it so successfully in so short a time—the reliance on innovative and cutting edge

The Satellite Technology Guide for the 21st Century

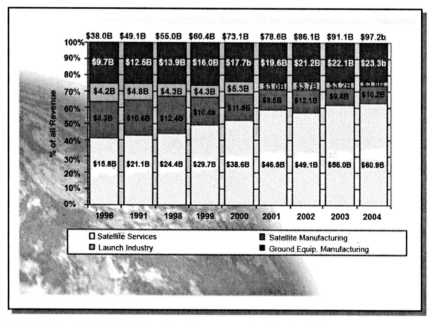

Table 2.1: Satellite Industry Revenues by Market Segment
Source: Satellite Industry Association

technology and the provision of reliable and dependable services.

In this chapter we will look at the composition of the satellite industry as it stands now in the middle of the first decade of the new millennium. It has come a long way from the monopoly of Intelsat in the beginning and has evolved into a major commercial industry worldwide.

Main Segments of the Satellite Industry

The satellite industry is composed of four main segments (see Table 2.2) namely: Satellite Manufacturing; Satellite Launch Service Providers; Satellite Services; and Ground Equipment Manufacturing sectors. All companies in the industry, over 9,000 of them according to the latest figures from the *International Satellite Directory* (2007), can be categorized to belong to at least one or two of these sectors. Some vertically integrated companies have subsidiaries active in all these four sectors. Most companies, however, specialize in just one area of the industry—for example some companies just manufacture satellites, but others such as Loral manufactures satellites through its subsidiary, Space Systems Loral, as well as own and operate satellites themselves with their

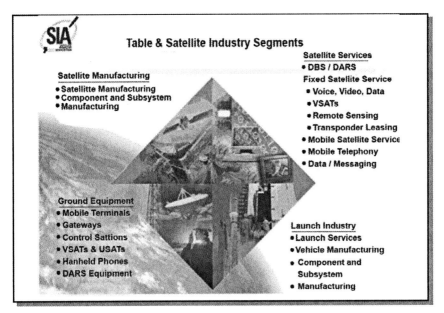

Table 2.2   Satellite Industry Segments
Source: Satellite Industry Association

subsidiary Loral Skynet (now merged with the Canadian satellite operator, Telesat).

Of all the segments of the satellite industry, the majority of the revenues (almost 60 percent in 2005) come from the provision and selling of satellite services. We will now discuss each segment and how they fit in the overall scheme of the satellite industry.

Satellite Manufacturing

The satellite industry began with the manufacturing sector—who took it upon themselves to turn Arthur C. Clarke's novel idea of providing worldwide satellite coverage into reality in the 50s and early 60s. The two pioneering companies in the field of satellite manufacturing discussed in the previous chapter are still in business today. Hughes Aircraft Corp. which developed and successfully launched the first geosynchronous satellite, Syncom-3 in 1964, was later renamed Hughes Communications Corporation and was bought by Boeing in the late 1990s and is now called Boeing Satellite Systems based in El Segundo, California. The company that launched the first passive satellite Echo-1 in 1960 and the first active commercial communications satellite, Telstar-1 in 1962,

The Satellite Technology Guide for the 21st Century

A typical satellite consists of the following major components seen here: solar panels for power; antennas for receiving and re-transmitting signals, transponders and its own propulsion system. (image of the WildBlue-1 broadband satellite courtesy of Space Systems Loral).

ATT's Bell Labs, is now Loral's Space Systems division called Space Systems Loral (SS/L) based in Palo Alto, California. Another U.S. company that manufactures satellites is Lockheed Martin Corporation. The U.S. companies are facing stiff competition from European manufacturers that have experienced considerable consolidation of its satellite manufacturing industry. The result of the process of consolidation among manufacturers in Europe are two formidable competitors: Thales Alenia Space and EADS Astrium (the result of the merger of the European Aeronautic Defense and Space Company and Astrium—which was formed from the merger of other satellite manufacturing companies that included Matra Marconi and Daimler Chrysler Aerospace).

Another noteworthy development in the satellite manufacturing sector is the rise of smaller manufacturers. In 2005, U.S.-based Orbital Sciences Corp. scored four contracts for the smaller satellites that they manufacture. Also, CAST of China, Melco of Japan, NPO M of Russia and Israel Aircraft Industries are increasing their share of the satellite manufacturing pie.

Basically, the satellite manufacturing sector produces the satellites that eventually get deployed in space. Satellites range in size from a basketball to a small car or bus (depending on the model of the satellite). A satellite is basically a communications system on its own with the ability to receive signals from the earth and to retransmit those signals back to the earth with the use of a *transponder*—an integrated receiver and transmitter of radio signals. The design and composition of a satellite will be discussed in greater detail in Chapter 4 "The Space Segment," but for the purpose of this discussion on the satellite industry as a whole—the main components of a satellite are the following: communications system (the antennas transponders that receive and retransmits signals); and the power and propulsion systems (the solar panels and jets that provide power and propels the satellite). The communications system of a satellite form part of the satellite that is called the "payload," while the power and propulsion system is part of what is called the satellite "bus."

The satellite manufacturing sector is not just composed of the aforementioned major manufacturers but rely on a host of other smaller companies and subcontractors that supply many of the subcomponents required in the assembly of a fully-functioning satellite. These consist of companies that provide electronic and communications components (i.e. antennas, transponders), and power systems such as solar panels and propulsion systems. Some companies specialize in providing just single components, while others provide a complete subsystem such as the communications or propulsion system of a satellite.

All these components and subsystems are assembled into the final satellite. Major satellite manufacturers brand their satellites under certain models or series of satellites, much like car manufacturers having several models or makes of their cars. For instance, Hughes has the famous HS-601 series of satellites, which have been rebranded as the Boeing 601 series. Space Systems Loral has its 1300 series, Lockheed Martin has the A-2100 series and EADS-Astrium, the E3000 satellite series.

Taking about 12-18 months or more to build, and costing upwards of US$200 million, manufacturers do not hold inventory of satellites ready for purchase. Satellites have to be ordered in advance as production depends on the schedule and backlog of the manufacturer. Being a "made-to-order" product, satellites can be customized according to the client's needs and specifications. The basic configuration can be based on an existing model of a satellite and provided with custom-built features.

The Satellite Technology Guide for the 21st Century

Satellites being manufactured today are a far cry from the first satellites in the early 60s that had just a single transponder capable of transmitting only one television channel. Satellites now have sophisticated electronics that can be configured to support various applications and thanks to advances in miniaturization and digitalization, they are now capable of receiving and retransmitting many orders of magnitude more signals than before. The latest Boeing 702 series of satellites, for example, can have up to 100 transponders. With the use of digital compression technology each transponder can have up to 16 channels, or a possible 3,000 or more broadcasting channels by one Boeing 702 satellite.

The result of advances in satellite technology has given rise to a healthy satellite services sector that provides various services to broadcasters, telecommunications and internet service providers, enterprises, government, military and other sectors.

Satellite Launch Sector

Once the satellite manufacturing process is completed and the satellite is fully tested, it is ready for launch into space. Satellites can be launched into different orbits: low-earth orbit—starting at about 100 miles from the earth; medium earth orbit (MEO)—starting at 1,000 miles from the earth; and geosynchronous orbit—22,300 miles from the earth (it is alternatively referred to as "geostationary" or "geosynchronous" orbit). The most popular earth orbit is the geostationary orbit, which was discovered by Arthur C. Clarke in 1945. At that precise distance from the earth, the satellite moves at the same speed as the earth's orbit, making it appear to remain stationary from the perspective of the earth's surface The big advantage of the geostationary orbit is that an antenna on earth can be fixed towards the satellite without it having to move or track the satellite's position. It also precludes the need for several antennas to track the satellites, as required in other orbits such as the LEO and MEO satellites. This makes geostationary satellites ideal for broadcasting and telecommunications services.

Launching a satellite requires a high-powered multi-stage rocket to propel it into the right orbit. This can be a very expensive proposition. Due to the high cost of the multi-stage rockets and rocket fuel required to launch a satellite, plus insurance against launch failure, satellite launches can cost between US$ 40-70 million for geostationary orbits and US$15-30 million for non-geostationary orbits.

In the past, satellite launches were also very risky. But the reli-

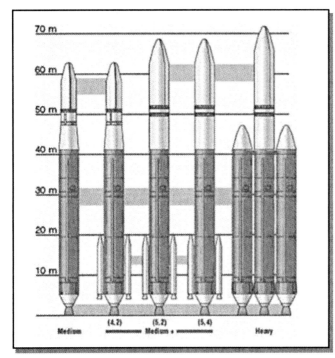

Boeing Delta IV family of launch vehicles—one of several models of launchers used to launch satellites into space orbit. (image courtesy of Boeing Launch Systems).

ability of launch vehicles has increased dramatically to about 94 percent in the new millennium, so we hear less and less of launch failures. The market leader for satellite launches is a French company called Arianespace which uses a launch facility in Kourou, French Guiana near the equator. Satellite launches are ideal at or near the equator because from that angle it would require a smaller trajectory to launch the satellite into orbit and therefore lesser fuel—a major part of the cost of the launch. Given that the launcher will have to travel a shorter distance from the equator to geostationary orbit, there is also lesser margin of error.

Other major satellite launch facilities are the Kennedy Space Center in Cape Canaveral, Florida and the Baikonur Cosmodrome in Baikonur, Kazakhstan. These two facilities figured prominently during the space race in the 60s and 70s between the US and the then-USSR. Satellites are also launched from the Vandenberg Air Force Base in California; Xichuan, China; and Tanagashima in Japan. Other launch facilities are being planned in Alcantara in the Amazon region of Brazil and in Christmas Island, Australia.

The satellite launch sector, like the manufacturing sector, is also

The Satellite Technology Guide for the 21st Century

dependent on various subcontractors to provide the launch vehicles—the rockets and electronics required to launch the satellites and other components and services related to the launch industry. Boeing is a leader in the production of rockets used for satellite launches with the Delta series of rockets. Lockheed Martin produces the Atlas series of rockets. Boeing and Lockheed Martin are also launch service providers and have recently banded together calling the joint-venture the "United Launch Alliance," which competes against the European launch service provider Arianespace and the Russian company called Energia, which uses the Proton series of rockets.

Arianespace manufactures its own launchers, the latest which is the Ariane 5 series. Rockets that launch satellites can be used for any object that needs to be lifted to space. The cargo carried by launchers are called "payload." The payload can be communications satellites or any cargo, or other vehicles including research and observation spacecraft as the Voyager series developed by NASA aimed at exploring outer space. Launchers can carry into orbit any payload and send it to space.

The US-based Orbital Sciences Corporation specializes in launching smaller satellites in non-geostationary orbit. It uses a smaller rocket called Pegasus which is piggybacked and launched from an aircraft such as the B-2 bomber. NASA's Space Shuttle also has the capability of launching satellites into orbit and has even performed repair and retrieval of malfunctioning satellites. However, due to the much-publicized failure of some of its missions, the Space Shuttle is considered the least reliable satellite launch option.

A new player in the launch industry is California-based Spacex, founded by internet entrepreneur Egon Musk (founder of Paypal). Spacex aims to provide alternative low cost launches of about $ 7 million (compared to the current going rate of $ 15-70 million) using its Falcon rocket.

Because of the advantages of launching near or at the equator, Boeing started an innovative company called Sea Launch based in Seal Beach, California. Sea Launch uses a seagoing launch pad to launch a satellite at sea where it will be closer to the equator. Since three-fourths of the earth's surface is covered in water, the most ideal sites for launching satellites are actually in the ocean as opposed to land.

Satellite Services Sector

Once the satellite is launched and placed in its correct orbital location—it goes through a series of in-orbit testing for a few months

Overview of the Satellite Communications Industry

Launch of the XM-1 Satellite for XM Satellite Radio from the Sea Launch platform in the Pacific ocean, May 8, 2001.
(image courtesy of Sea Launch).

before it is deemed ready for service. Once operational, a satellite has the capability of providing a variety of services. Digitalization, which reduces all information to bits of 1s and 0s, makes it possible to receive and transmit data, voice, audio and video all in one transmission stream.

Satellites perform a wide variety of indispensable services to society. Without satellites, we will not be able to receive most phone calls, television and cable programs and a great deal of the vital data that makes our information-driven society work.

Naturally, the capability to receive and retransmit signals to a large coverage area over vast distances has its advantages. Satellites first made its impact by making telephone calls possible across the Atlantic and other oceans. Later satellites made possible the transmission of live events worldwide in real-time. It used to be standard for satellite broadcasts to carry the "Live Via Satellite" tag, but now satellite delivery of programming has become so ubiquitous that broadcasters have stopped doing so. We now take for granted that a great majority of programming that we receive go through satellites at some point or another.

The Satellite Technology Guide for the 21st Century

The various services that satellites provide will be discussed in detail in Chapter 6 "Satellite Services." In general, there are three types of services that geostationary satellites provide: telecommunications; broadcasting and data communications. Other satellite services include remote sensing, imagery, navigation, meteorological or weather monitoring, scientific research and military applications.

Telecommunications services include telephone calls and other services provide to the telecommunications sector which include telephone companies, wireless and mobile telephony providers, and PTTs (Post, Telephone and Telegraph) authorities.

Broadcasting services include radio, television and Direct-to-Home services and mobile broadcasting. Direct-to-Home (DTH) are satellite television services that are received directly by households such as the DirecTV and Echostar services in the US. Now even radio programs can be broadcasted directly by satellites. Soon, satellites will also play an important role in delivering programming to cell phones and other mobile devices such as personal digital assistants (PDAs), laptops and other handheld technologies.

Data communications involve the transfer to data from one point to another. A lot of corporations and organizations require massive amounts of data such as financial and other information exchanged between its various locations. The ideal manner by which data and other information is exchanged among companies with locations such as a retail chain, for example, is through a technology called VSATs or very small aperture terminals. Some companies specialize in provision of VSAT-based services like Hughes Network Systems and Gilat Satellite Networks, among others. With the growth of the internet, a significant number of internet traffic goes through satellites, making Internet Service Providers (ISPs) one of the largest customers for satellite services.

In the provision of services, satellites face competition from other media such as fiber, cable and other land-based delivery systems such as microwaves and even powerlines. The main advantage of satellites is that it can distribute signals from one-point to many locations. As such, satellite technology is ideal for "point-to-multipoint" communications. Satellites also do not require massive investments on the ground—making it ideal for underserved areas like isolated rural areas.

Satellites and other delivery mechanisms such as fiber and cable are not mutually exclusive. In the design of communications networks, one may need a combination of various delivery mechanisms. This has given rise to various hybrid solutions, where satellites form one part of the value chain in combination with other media.

The main providers of satellite services are satellite operators—those who own and operate their own satellites. Satellite operators are dominated by a few that have satellites in key orbital positions worldwide and therefore can provide global coverage. Some operators provide only regional coverage, such as only in Latin America, for example, or some only provide coverage to one national area, like some of the satellites covering only Russia or the former Eastern Bloc countries. The advantage of having global coverage is that you can seamlessly provide services using your own satellites throughout the world. This is valued by clients who need global coverage such as some satellite broadcasters and telecommunications companies.

There are also companies that specialize in providing satellite services, without owning any particular satellite. These companies usually have ground facilities of their own and lease satellite transponder space from satellite operators. These service providers usually offer an "end-to-end" solution to its customers using facilities culled from various sources. Some of the largest satellite service providers include France-based GlobeCast, US-based Ascent Media, the UK-based Arquiva, among others. The large service providers operate their own facilities called "Teleports" which have the capability to originate programming, receive and transmit signals from satellites and also provide connectivity with other terrestrial networks such as connections to the fiber network. Service providers, however, run the gamut from small one-person operations to large companies with various facilities and multi-transponder leases. Basically, any entity that deals in satellite services—as an originator or operator of the service or a reseller or broker of satellite services is part of the satellite services sector.

Ground Equipment Sector

The provision of satellite services requires various equipment on the ground to originate, receive and transmit signals to satellites. The equipment required by satellites service providers on the ground will be discussed in detail in Chapter 5 "The Ground Segment."

Basically, the ground equipment sector provides the various uplink and downlink equipment necessary to communicate with satellites up in orbit. Some of these equipment consist of Telemetry, Tracking and Control (TT&C) equipment that monitors and tracks the satellite's posi-

Table 2.3 Top Satellite Operators
(in terms of number of satellites in operation)

RANK	COMPANY (Country)	2006 REVENUES (US $)	SATELLITES IN OPERATION (as of Dec. 2007)	NOTES
1	INTELSAT (Bermuda)	1.7 Bil.	51	Merged with PanAmSat in 2006
2	SES Global (Luxembourg)	1.9 Bil.	36	Acquired News Skies Satellites in 2006
3	Eutelsat (France)	880 Mil.	19	Owns 27% of Spanish operator Hispasat
4	TELESAT (Canada)	575 Mil.	12	Merged with Loral Skynet (USA) in 2007
5	Russian Satellite Communication Co. (Russia)	152 Mil.	11	Partnership with Intersputnik in several satellites
6	Indian Space Research (India)	76 Mil.	10	Organization owned by the Indian government-operates a global marketing arm as Antrix Corporation
7	JSAT (Japan)	349 Mil.	8	Owns 50% of Horizons satellites with Intelsat
8	APT Satellite Holdings (Hong Kong)	54 Mil.	5	Primarily owned by Mainland Chinese companies
9	ArabSat (Saudi Arabia)	150 Mil.	5	A regional organization founded in 1976 by 21 member-countries of the Arab League
10	SingTel Optus (Australia)	192 Mil.	4	A merger of Singaporean and Australian telecom companies. Its parent, Singapore Telecom operates independently one satellite, ST-1, and co-owns and leases transponders on APT satellites

tion. There are also uplinking equipment that converts the signal and transmits it to the satellite like upconverters, modulators, amplifiers and antennas. Downlink equipment which receives signals from satellites includes antennas, demodulators and downconverters.

The ground equipment sector also produces equipment for the VSAT market and the consumer satellite services such as Direct-to-Home (DTH), Satellite Radio and Internet via Satellite. Among the various sectors of the industry, the ground equipment market is probably where the most growth can be seen in the long-run. Comprising only 28 percent of overall industry revenues, the relatively small share of the ground equipment market in terms of industry-wide revenues will only grow as the demand for satellite services increase. As the adoption of consumer satellite services pickup in various regions of the world, the demand for satellite equipment will only grow. According to analysts, broadband satellite services, which are starting to take off in the US with the WildBlue system and in Asia with IPStar service, which enables one to receive broadband internet using a small dish and terminal, will continue to grow in the coming years. Once the consumer sector takes off, demand for satellite equipment will run into the millions of units as opposed to the more specialized equipment for satellite service

A typical ground equipment to receive satellite signals is WildBlue Communications' consumer 29.1-inch antenna and satellite modem which connects to a PC and can receive and transmit broadband internet. (photo courtesy of WildBlue Communicatons).

There are over 40 satellite companies operating geostationary satellites in the world today. Of these, only four (the top 4 in the chart in the previous page) have true global coverage. The rest have regional or only national coverage. The top four companies with global coverage control an estimated 60 percent of revenues of the entire satellite operators' sector of the industry.

Data culled from company sources by Satellite Markets and Research (www.satellitemarkets.com)

providers which have a smaller manufacturing run.

Jobs in the Satellite Industry

An estimated 60,000 people directly work in the satellite industry worldwide in 2007. Many more work in industries related or allied to the satellite industry, such as the cable, broadcasting, telecom and internet industries. A misconception on the satellite industry is that it is overwhelmingly composed of engineers and technical people. However, engineers and those with technical backgrounds compose probably less than one-third of the industry. Although having a technical background is a plus in the industry, many have had successful careers in the satellite industry without a technical degree. Since the industry is constantly changing and evolving, longevity in the industry requires more of an aptitude to learn and adapt to changing situations than a rigid technical training.

Like many industries, the satellite industry needs sales and marketing personnel which could come from any background, although as previously mentioned those with a technical background would have a definite advantage. Many a company in the satellite industry has failed with a great technical product backed by poor marketing acumen. Intense competition and little differentiation in satellite products make savvy marketing an important ingredient for success in the satellite industry. The sales and marketing department of a satellite company can make or break a company.

Satellite companies also need proven corporate executives in various areas: legal, finance, public relations, customer service, human resources and training. Project managers are particularly valued for managing complex satellite projects.

There are also many other jobs in sectors following the satellite industry as analysts, researchers, consultants and writers. The influx of private investment in the satellite industry has opened up new opportunities in this field. There are also those who work in government regulatory agencies and non-profit organizations and media enterprises such as Satellite Markets and Research (www.satellitemarkets.com) that monitor the industry.

As far as I know, there is no university degree specifically focused on satellite technology as such. Satellite technology is usually a specialization under engineering or communications programs. Much of it is learned on the job or in continuing education such as attendance in seminars, conferences and trade shows. So, don't let the lack of a technical background hold you up from making a career out of the exciting field of satellite communications.

Very few industries provide the cutting edge technology and the opportunities and challenges such as in the satellite industry. Moreover, the satellite industry provides an indispensable social service that profoundly affects almost every aspect of modern life.

CHAPTER 3

The Basics

To be able to understand satellite communications and how it works, it is necessary to have a basic knowledge of the underlying scientific principles governing the technology. Satellites are a type of communications medium that receive signals from a transmission source and retransmit the signals to a receiver or a number of receivers. The basic underlying property governing satellite communications are radio waves, a force inherent in nature just like gravity and radiation. Satellite communications is made possible by the transmission (or communication) of a signal from a source (or sources) to a receiver or many receivers. As a communication process, satellite communications follows the same model that governs all types of communications, whether interpersonal or electronic communications. Communications always occurs from a given source transmitting a signal through a medium which is received at the end by a receiver. Communication theorists have called this model the SMCRE model of the communication process or the Source-Message-Channel-Receiver-Effect model. A simple diagram of this model is shown in Figure 3.1 below:

Figure 3.1: The SMCRE Model of Communications

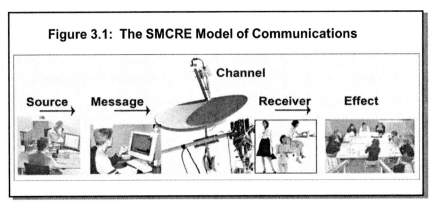

The Satellite Technology Guide for the 21st Century

As mentioned earlier, all forms of communications, from the simple conversation you had with your friend standing right next to you, to telephone conversations across the Atlantic via satellite links follow this simple model. Each one has a source (the initiator of the conversation) a message (the words being spoken); a channel (in the telephone conversation illustrated in Figure 3.1 —it uses several channels or media—including electrical pulses through telephone wires, then through microwave transmission to the satellite); and the message is received by the other party in the conversation producing an effect which could be feedback and therefore a repetition of the same process.

Of course, communications can be very complex, using many different media and can have many sources trying to send messages at the same time. There are also atmospheric and other terrestrial factors (even the temperature affects the signal of radio transmissions) that could interfere with communications. These are generally called "noise" or "interference." The less noise and interference the better the intended message will be received by the receiver and the higher fidelity of the signal from its source. So an added dimension to the model of communi-

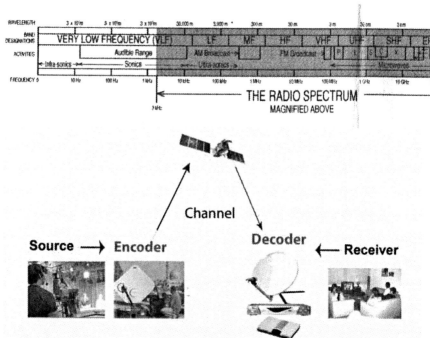

Figure 3.2: Source-Encoder-Channel-Decoder-Receiver Model of Communications

The Basics

cations is the existence of noise. One way of minimizing outside noise and interference, and to ensure that the message only goes to the intended receiver and no one else is to encode the message. So in satellite communications, there is an added step in the communications process that encodes the message before transmission through the channel and is decoded at the other end. A modified communication model for electronic communications as propounded by Claude Shannon who formulated the Information Theory that governs new media technologies including satellites, will look like the Source-Encoder-Channel-Decoder-Receiver Model of Communication in Figure 3.2.

In this model, a source, which could be the video and audio from a broadcasting studio, which in engineering parlance is called the *baseband* signal, is encoded for transmission through a channel—the satellite. The baseband signal is encoded or modulated according to the frequency of the satellite. This modulated signal is then downlinked from the satellite by a receiver and decoded or demodulated back to its baseband form so that it can be viewed accordingly. These are the underlying process that involves all radio communications and an explanation of each of these important concepts follows.

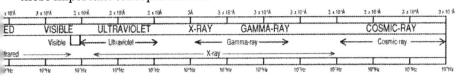

Figure 3.3. The Electromagnetic Spectrum which range from very low frequency waves to cosmic waves. The radio frequency bands through which satellite communications are transmitted form only a small part of the whole spectrum. (Source: US Department of Commerce)

Satellite technology, is made possible by the transmission of radio waves. Radio waves are invisible particles that can travel very long distances at the speed of light or 186,000 miles per second. They have electronic and magnetic properties and that's why they are part of the electromagnetic spectrum, which as previously mentioned is one of the major naturally occurring forces in nature (gravitation and radiation are the others). Humankind has been able to harness electromagnetic waves since its properties were discovered by a Scottish scientist named James Maxwell in 1864. A German scientist named Heinrich Hertz discovered

Radio Frequency Bands

Medium Frequency (MF) Bands: AM Radio (535-1705 kHz)

Shortwave Radio (1.6-30 MHz)

Citizens' Band Radio (27 MHz)

Very High Frequency (VHF) Bands:

FM Radio (88-108 MHz)

VHF Television (54-216 MHz)

Super High Frequency (SHF) Bands:

Satellite Communications (1-30 GHz)

that radio waves form part of a larger electromagnetic spectrum which includes infrared light and later it was found out also includes X-rays (see Figure 3.3). For his contribution to the understanding of radio technology the measurement of the frequency of magnetic waves are measured in Hertz (Hz).

Frequency is the number of oscillations of a given wave. Each frequency is one oscillation per second measured in Hertz. So one Hertz is one oscillation (or vibration) of the wave per second as it passes through

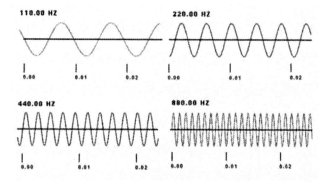

Figure 3.4: Frequency is the measurement of the number of vibrations per second of an electromagnetic wave.

The Basics

one point. Just as different vibrations produce different pitches of sound, different frequencies can be used to transmit distinct signals that are distinguishable by a receiver tuned to the same frequency. This is the origin of the expression being on the "same wavelength "or "being on the same frequency." A portable radio is essentially a tuner that can "tune in" to various frequencies i.e. AM or FM radio signals. Satellites operate on much higher frequencies and do not interfere with terrestrial AM or FM broadcasts.

Now that we know that almost all electronic means of transmission involves the use of radio or electromagnetic waves, let's look at the most common terms associated with radio technology of which satellite communications is an integral part of.

Orbital Position

The most basic term used to identify satellites apart from their corporate name, for example Intelsat's IS-2 satellite covering the Pacific region, is its orbital position, which is 169^0 E (the "E" stands for East). 169^0 E denotes its position in the geostationary orbital arc. Since most

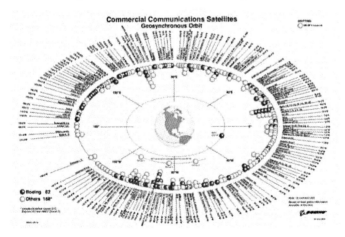

Figure 3.5. As can be seen from this graphic representing the number of satellites parked in the limited orbital slots in the geostiatonary arc—some regions are saturated, such as North America and Europe, leaving only limited slots available, while certain regions such as Africa are underserved. To maximize the use of orbital slots, several satellites can be parked in the same orbital position operating in different frequencies to avoid interfering with each other. (Source: Boeing Satellite Systems).

41

The Satellite Technology Guide for the 21st Century

commercial satellites are in the geostationary arc which rings around the earth—the whole arc around the equator is a perfect 360^0. The arc begins at the Greenwich Prime Meridian (near London) which is 0^0 longitude. The orbital arc is divided into 180^0 west and 180^0 east of the prime meridian. The orbital position of the satellite determines the elevation angle from earth stations to the satellite, and the coverage and utility of a transponder for a given network application. Like any business, location is pivotal and this is true in the case of securing key orbital slots.

The body that regulates the orbital arc is the International Telecommunications Union (ITU) based in Geneva, Switzerland. The ITU receives and approves applications for use of an orbital slot for satellites. Because orbital slots are a very finite resource, with only 180 slots available at any given time (since the ITU mandates 2 degrees spacing between satellites to avoid potential interference), the criteria for granting orbital slots are very stringent. Theoretically, any entity, public or private, can apply for orbital slots—but they have to meet the ITU's rigid criteria including payment of a hefty deposit and other fees. There are severe penalties for not meeting ITU requirements, which can include a

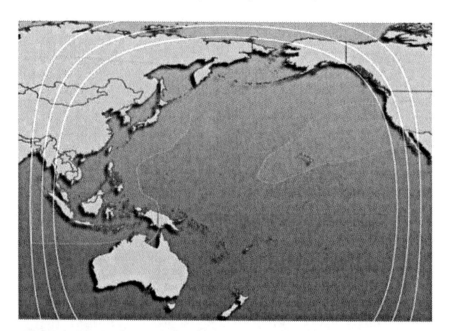

Figure 3.6 Footprint map of IS-2 satellite covering East Asia, the Pacific and parts of the West Coast of North America. Note the different coverage "beams." (Source: Intelsat)

The Basics

provision to launch a satellite in orbit within two years of obtaining the license to operate in a specific orbital position. Due to the high demand for certain regions, almost all available slots are already taken in the heavy traffic areas of North America and Western Europe.

One way of maximizing an orbital slot is to park several satellites in the same orbital slot. As many as seven SES ASTRA satellites, for example, are sharing one orbital slot at 19.2^0 E in the crowded Western Europe area. They are able to share one slot without interfering with one another by using different frequency bands to receive and transmit signals.

Satellite Footprint

The coverage area of a satellite, or the area in which it is possible to receive the signals from a specific satellite is called the satellite's "footprint." Satellite companies provide footprint maps of their satellites in order to show the relative power in every area of its footprint. A satellite's signal cannot be received outside of its footprint. It can, however, be received by an earth station in its footprint and reuplinked to another satellite that has coverage in another area. For example in Figure 3.5, we

\multicolumn{3}{c}{SATELLITE FREQUENCY BANDS}		
Band	Frequency Range	Services
L	1-2 GHz	Mobile Services
S	2.5-4 GHz	Mobile Services
C	3.7-8 GHz	Fixed Services
X	7.25-12 GHz	Military
Ku	12-18 GHz	Fixed Services
Ka	18-30.4 GHz	Fixed Services
V	37.5-50.2 GHz	To be Decided

can see the footprint of Intelsat's IS-2 satellite which covers the Asia-Pacific region. In order for a receiver in the Eastern part of the US, which is not in the satellite's footprint, to get a signal from IS-2 an earth station in its footprint, say in Hawaii, can downlink the signal, and re-uplink to a satellite that covers the Eastern US that can be seen from Hawaii. Due to the number of satellites in the geostationary arc, many footprints of satellites overlap, thereby allowing satellite "hops" as described in the aforementioned example. Since some satellites cover as much as one-third of the earth, it is possible to carry a signal around the world with three "hops" from different satellites.

As can be seen from the footprint map—a satellite's coverage can have different "beams" or a satellite's transponder's signal focused on a relatively smaller area. One can choose to transmit to a large area which can be an entire continent or to a much more concentrated area such as specific regions.

Frequency

As mentioned earlier, various forms of radio communications operate in different frequencies. Frequencies are such an integral part of radio communications that satellite engineers, like their counterparts in radio broadcasting, are called "RF" engineers, which stands for "radio frequency." Transmissions operate in various frequencies in order to differentiate signals (otherwise everybody will be transmitting on the same frequency and there will be massive interference rendering communication impossible). Each operating frequency has different advantages and disadvantages. Terrestrial radio is broadcasted in the AM or FM frequency bands while television is broadcasted in the VHF and UHF frequency bands.

Satellite communications occur in the very high frequency range of 1-50 GHz (one GHz=one billion Hz) the frequency ranges or bands are identified by letters such as C-, Ku-, Ka-, S-, X-, and V-Bands. The lower range of the spectrum C-band signals have lesser power and need larger antennas to receive their signals.The higher end of this spectrum, the Ku- and Ka-Bands have more power and therefore need a smaller dish to receive the signal (as small as 60 cm or 24 inches—about the size of a medium pizza). This makes the Ku-band and Ka-band spectrum ideal for applications such as Direct-to-Home (DTH) broadcasting and two-way broadband communications.

A satellite operates in various frequencies in its assigned range.

Each transponder (the key element of a satellite that receives and retransmits signals) is assigned a frequency to operate in. One transponder is usually assigned to one signal which can be an analog signal or digital which can be shared by several channels. An example of an analog signal is one television channel using a 36 MHz transponder for its transmission or a digital signal that compresses the signal enabling up to 16 television channels to share one transponder.

The ITU regulates and assigns the different bands for specific applications for each region. Every few years, the ITU convenes the World Administrative Radio Conference (WARC) which is responsible for assigning frequencies to various applications in various regions of the world. Each country's telecommunications regulatory agency enforces these regulations and awards licenses to users of various frequencies. In the US, the regulatory body that governs frequency allocation and licensing is the Federal Communications Commission (FCC).

Satellite Frequency Bands

C-Band

C-Band is the most widely used frequency for satellite communications around the world. In North America, C-Band is used primarily for cable TV programs, broadcast network signal distribution, and radio distribution. The network "feed" or program signal is beamed over satellite from the network origination point to receive antennas located at affiliated TV stations or to cable TV systems. Moreover, international distribution of television and radio signals to overseas cable, broadcast, and Direct-to-Home (DTH) centers is performed in many parts of the world using C-Band transponders on satellites with coverage designed to cover multiple countries and regions.

C-Band satellite transponders, such as Intelsat's "Global Beam" transponders, are capable of covering very large areas – as much as 40 percent of the earth's surface.

As a tradeoff for coverage size, C-Band transmissions require larger ground antennas than Ku-band transmission due to the lower frequency of C-Band and the larger radio wavelength. C-band Television Receive Only (TVRO) antennas are not typically smaller than 2.4m (7 feet) in diameter.

The Satellite Technology Guide for the 21st Century

Ku-Band

Ku-Band allows for use of smaller receive antennas, due to the higher frequency, shorter wavelength, and typically higher satellite transponder power. In North America, receive antennas at Ku-Band can be as small as 60-90 cm (2-3 feet), although the smaller the dish, the more likely the risk of interference from neighboring satellites in what is called the conventional Ku-Band. This problem does not apply to true DBS (Direct Broadcast Satellite) satellites in the Broadcast Satellite Service band, which are spaced farther apart in orbit and work well with 18-inch antennas in the US.

In parts of the world where C-Band frequencies are used extensively for terrestrial microwave communications, such as throughout much of Western Europe, radio frequency interference can be a challenge in setting up C-Band networks particularly near cities. As a result, in these regions, Ku-Band satellite transponders are in greatest demand and use.

Ka-Band, at higher frequency frequencies (27.5-31.0GHz uplink/

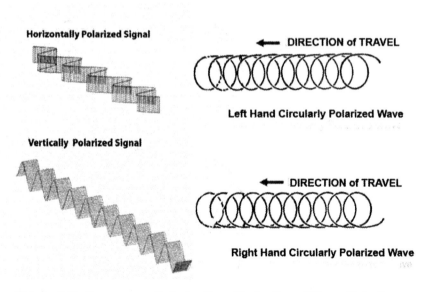

Figure 3.7: Frequency Polarization. To further differentiate frequencies, they can have linear or circular polarization. Polarization indicates the direction the frequency wave travels. A satellite transponder, for example can have two different polarizations of the same frequency, maximizing frequency use.

AMC-1 Satellite Frequency Plan

Transponder	UPLINK	DOWNLINK	POLARIZATION
1	5945	3720	H/V
2	5965	3740	V/H
3	5985	3760	H/V
4	6005	3780	V/H
5	6025	3800	H/V
6	6045	3820	V/H
7	6065	3840	H/V
8	6085	3860	V/H
9	6105	3880	H/V
10	6125	3900	V/H
11	6145	3920	H/V
12	6165	3940	V/H
13	6185	3960	H/V
14	6205	3980	V/H
15	6225	4000	H/V
16	6245	4020	V/H
17	6265	4040	H/V
18	6285	4060	V/H
19	6305	4080	H/V
20	6325	4100	V/H
21	6345	4120	H/V
22	6365	4140	V/H
23	6385	4160	H/V
24	6405	4180	V/H

Satellite Frequency Plan of SES Americom's AMC-1 satellite showing each transponder's uplink and downlink frequency and polarization. (Source: SES Americom)

18.3-20.2 GHz downlink) than Ku-Band, offered two promising benefits:

• A large increase in the amount of available spectrum (3500MHz) compared with Ku and C-Band (500 MHz per polarity, per satellite position in an increasingly crowded spectrum).

• The potential to deliver high bit rates into smaller antennas on the ground due to the increased bandwidth and frequency.

The Satellite Technology Guide for the 21st Century

A technical challenge with Ka-Band is the large impact of rain attenuation on signal availability. Concerns about the impact of rain fade on Ku-band satellite links were a perceived drawback when Ku-Band was first introduced as a new source of spectrum, but over time improved system performance and operational experience has shown that Ku-band rain fade, although much higher than C-Band, is acceptable for most applications.

Ka-Band satellites have been used in commercial test and demonstration payloads since the 1990s (Italsat 1, Japan's Superbird and N-star). NASA's ACTS (Advance Communications Technology Satellite) program demonstrated the feasibility of Ka-Band communications applications, using smaller ground antennas than Ku-Band dishes in 1993. In 2005, Ka-Band broadband satellite services were launched in Asia by Thai operator Shin Satellite with its IPStar service and in the US with WildBlue Communications. Hughes Network Systems, a leading provider of satellite delivered broadband services in the US, launched in 2007 a Ka-Band satellite, SpaceWay-3, to serve their new customers with Ka-Band delivered broadband service.

Polarization

Apart from having different frequencies, radio waves can have different polarization, or the property of electromagnetic waves that describes the direction of the transverse electric field. Signals can be linearly polarized, either Horizontal or Vertical polarization, or it can be circularly polarized, which can either be left-hand or right hand polarization.

Polarization of a signal is just another way of differentiating signals and also maximizing use of available frequencies. Thus a satellite transponder operating in 6.305 GHz C-Band frequency, for example, can have a horizontal or vertical frequency operating on the same frequency or if it's a circularly polarized satellite it can have left- or right-hand polarization. This is illustrated by the example in the previous page of a Frequency Plan of a satellite. Frequency plans are vital information on a satellite that is publicly available in most satellite operators' websites or with commercial databases. As can be seen from the frequency plan, the downlink (the signal received from the ground by the satellite's transponder) and uplink frequencies (the signal retransmitted to the ground by the satellite's transponder) can either be horizontal or vertically polarized. This way, two signals can share the same uplink or downlink frequency without interfering with each other.

The Basics

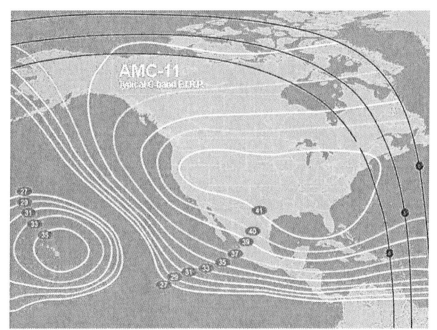

Figure 3.8 EIRP Map of SES Americom's AMC-11 satellite which covers North and Central America and Hawaii. The higher the dBw, in this case, 41 dBw, the higher the power of the signal from the satellite. (Source: SES Americom).

Power

Another important term that you cannot avoid in satellite communications is Decibels. You may have learned in basic science that sound is measured in Decibels. Decibel is named after Alexander Graham Bell—the inventor of the telephone. The power of the signal from satellites transmissions are measured in Decibels per one watt of power or dBw. You will see this in EIRP maps, another common piece of information on satellites which show the relative power of the signal of the satellites relative to a position on the ground. EIRP stand for Effective Isotrophic Radiating Power and can vary from a low of 27 dBw to a high of 40+ dBw for C-Band satellites. Ku-band satellite transponders can have dBws in the 50s. The higher the dBw—the higher the power of the signal, meaning a smaller dish can be used to receive it on the ground. To illustrate the increase in power represented by a higher dBw—each increment of 3 dBw roughly represents a doubling of power. So a signal with 50 DBw is roughly 100,000 times stronger than a signal with 3 dBw.

The Satellite Technology Guide for the 21st Century

The EIRP of a satellite is very important piece of information as it will determine the type and power of ground receiving equipment and is a vital part of the equation call "Link Budget Analysis" which RF engineers use to calculate the signal strength versus noise in the link from the sending point ("uplink") and over the satellite to the receive point ("downlink") point. The link budget depends on the following prameters:

- Uplinked signal parameters and power
- Transponder downlink power (EIRP)
- Receive antenna performance
- Signal modulation parameters
- Interferences and "noise" in the transmission chain.

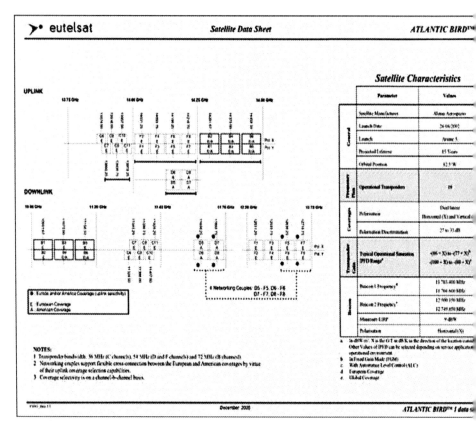

Figure 3.9: An example of a typical satellite data sheet with the frequency plan and satellite specifications detailed. (source: Eutelsat).

50

The Basics

If you are not an engineer, it's unlikely that you would be ever required to perform a link budget analysis. However, it's important that you understand the process. There are several user-friendly Link Budget soft ware available such as Satnews Publishers' Satfinder and Arrowe Technical Services' Satmaster Pro.

Now that you can read an EIRP map and a frequency plan, two very important sources of information on satellites, let's discuss other key parameters that often come to play in evaluating a satellite.

Signal to Noise ratio (S/N)

Signal to Noise (S/N) ratio is a measure of signal strength relative to background noise. The ratio is usually measured in decibels (dB). For analog signals, the ratio, denoted *S/N*, is usually stated in terms of the relative amounts of electrical power contained in the signal and noise. For digital signals the ratio is defined as the amount of energy in the signal per bit of information carried by the signal, relative to the amount of noise power per hertz of signal bandwidth. RF engineers always strive to maximize the S/N ratio of a signal, which will provide a higher quality of signal to receivers.

Gain/Temperature

Gain/Temperature or G/T is an important parameter in satellites that measures the ratio between the "Gain" of the system, which is the net between the signal output to the signal input of a system, and the "temperature" or the amount of noise in the system. The higher the G/T ratio the greater the fidelity of the signal. G/T has wide applications in amplifiers in antenna systems, which are use to amplify weak satellite signals.

SFD

SFD stands for saturation flux density which is the power required to achieve saturation of a single channel on a satellite. Saturation is reached when an amplifier reaches the non-linear part of it power transfer, such that an increase in input power results in little or no increase in output power.

Analog vs. Digital

Satellites are able to receive and retransmit four different types

The Satellite Technology Guide for the 21st Century

of signals: video, audio, data and telephony. These signals can either be analog or digital. Analog signals are baseband signals that are transmitted as electromagnetic waves with varying frequencies. Digital signals on the other hand are transmitted in the form of binary bits of 1s and 0s. Because there are only two possible values in the binary system—either *on* (represent by ones) or *off* (represented by zeros) —digital signals are clearer and less susceptible to noise and interference than analog signals. Digital transmissions are also generally higher speed than analog transmissions.

Although both analog and digital signals are subject to degradation—digital signals can be regenerated, while analog signals cannot. Apart from being a lot clearer than analog signals, the great advantage of digital signals is that the can be compressed and decompressed with little degradation or loss of the signal—which is an important consideration when you have limited bandwidth. With transponder leases on satellites costing over a million dollars a year—digital compression enables several channels to share one transponder. So instead of one TV channel, for instance, using a 36 MHz transponder on what is called a Single Channel Per Carrier (SCPC) system, you can have Multiple Channels Per Carrier (MCPC) with ratios of from 1:2 (one transponder shared by two video channels) to as high as 1:16 (even more for audio channels).

Analog signals can be converted to digital and vice-versa using analog-to-digital converters (ADC) or digital-to-analog converters (DAC).

Bandwidth and Bit rates

Bandwidth is the measure of the frequency use or capacity. The higher the bandwidth the greater its capacity to carry signals.

In the digital realm, capacity is measured in bits per second (bps). The capacity of a digital-capable satellite transponder is measured in Megabits per second (Mbs).Satellite transponder capacity depends on factors such as the type of digital compression used and the link budget performance.

An analog TV signal typically occupies from 18 to 36 MHz, or a full transponder's bandwidth, while professional broadcast video that is digitally compressed using today's available MPEG technology can use a fraction of the transponder's bandwidth. MPEG stands for the Motion Pictures Expert Group which develops video compression standards. Another group that develops video broadcast standards is the Europe-based DVB Project, an industry consortium with more than 270 members, which promotes the Digital Video Broadcasting (DVB) standard. While most digital video signals today are encoded in MPEG-2, the introduction of new video compression technology like MPEG-4 promises to add more video signal compression and reduce the bandwidth required to transmit video.

There are many types of digital compression techniques including Binary Phase Shift Key (BPSK), Quadrature Phase Shift Key (QPSK), 16-QAM and 32-QAM (Quadrature Amplitude Modulation). What is important to understand is that both analog and digital signals can be compressed and encrypted to control who should be able to receive the signals. Signals can therefore be encrypted using different techniques, which requires a special receiver to decode the signal on the ground called an IRD or "Integrated Receiver-Decoder. "A signal can also be unencrypted and can be received by anyone with basic reception equipment. Unencrypted signals are called "free-to-air" or "in-the-clear" signals. Some satellites provide information on what signals it transmits and this information is collected by companies such as SatcoDX (www.satcodx.com). An excerpt of a typical listing of what's on a satellite would look something like this:

Transponder no.	Frequency	Polarization	Programming	Format
1A	3.677	H	Discovery	DVB
1B	3.860	V	MTV	MPEG
2B	4.180	V	Global TV	in the clear

As more and more bandwidth-intensive applications are developing in the market like High Definition Television and IPTV, for instance, satellite manufacturers are keeping in step with the demand by designing larger satellites with more transponders and higher power.

Basic Satellite Network Architecture

Now that you are familiar with the basic concepts of satellite transmission, let's look at the most basic elements of a satellite communications system. A satellite network has two main components: the space segment which consists of the satellite spacecraft itself and the ground segment which consists of the transmission and reception equip-

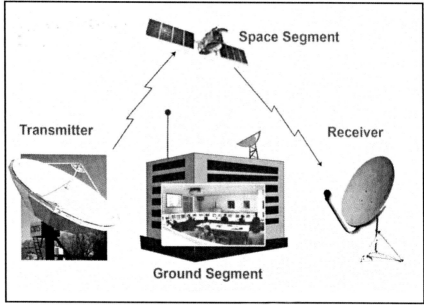

Figure 3.10. Typical Satellite Network.

ment on the ground, illustrated by Figure 3.10.

In the next two chapters, we will discuss in detail the Space Segment and the Ground Segment respectively.

CHAPTER 4

The Space Segment

The satellite spacecraft is really the heart and soul of a satellite network, and one can even say of the satellite industry. The satellite industry as we know it today was made possible by the pioneering work in the 50s and 60s by engineers in satellite manufacturing companies, namely Hughes Aircraft now Boeing Satellite Systems and AT&T's Bell Labs which is now Space Systems Loral. The current satellite models are a far cry from their early predecessors just a little over 40 years ago. Not only are they much bigger, they are more powerful and can perform so much more functions that enable many of the new emerging applications.

Today's satellites are truly a marvel of engineering and design. They are the product of many years of technical innovation and the best that the experiences and lessons learned from Space Race in the 50s and 60s and the space industry that it engendered.

A note on terminology—man-made satellites are also called spacecraft, or more specifically, satellite spacecraft. They are also referred to as "artificial satellites" to differentiate it from natural satellites that are heavenly bodies like the moon which orbit around a planet. Informally, satellites are also referred to as "birds" after the first Intelsat satellite "Early Bird." In this book, we will just refer to satellite spacecraft as "satellites."

Because it has to operate in the harsh environment of space without any prospect of physical repair by a live technician during its projected life of 10-15 years, a satellite has to be precisely engineered and made of very durable materials. A satellite in space should be able to withstand the shock of a rocket launch where it is thrust into space at 17,500 miles per hour causing tremendous vibrations. When placed in orbit, the satellite has to be able to endure extremes of cold and hot temperatures which could swing 200^0 C in either direction. To complicate this further, satellites have to be lightweight as the cost of launching a satellite can be up to US$ 15,000 per kilogram (kg) (or US $ 60 million for a 4,000 kg satellite). To meet all these challenges satellites are made of lightweight and durable materials compressed in a very small space, and still manage to operate at very high reliability of over 99.9 percent in the vacuum of space with no prospect of maintenance or repair during its entire lifetime.

Despite the limitations of its design—its compact size, its use of lightweight materials and miniaturized equipment, and bereft of a live human operator—a satellite, about the size of a small car, performs essentially the same functions as a medium-size earth station or teleport with much more equipment occupying thousands of square feet of space and operated by a full-time staff.

A geostationary satellite from the major manufacturers can cost over US$ 200 million dollars and can take from two to three years to build. Thus, each geostationary satellite is made to order. There is no inventory of ready-made satellites you can buy off-the-shelf nor are there any used satellites for sale. Once a satellite is launched in to space, it is pretty much irrecoverable (the satellite just ends up as space debris after its 10-15 year lifespan). The Space Shuttle has successfully retrieved satellites that malfunctioned in the past, and are able to repair and reposition satellites to another orbital location, but that is a very expensive proposition. So, pretty much the only option for anyone planning to procure a satellite is to order a new one.

Due to the high cost entry into the satellite business, very few satellites are made every year. In 2006 only 21 geosynchronous satellites were ordered worldwide from the major satellite manufacturing companies (coincidentally 21 satellites were also launched into geostationary orbit in the same period).

Satellite design is very customer-driven. If you happen to have the resources to procure a satellite and pass the initial screening and credit checks by the major satellite manufacturers, you will probably be handed a voluminous questionnaire about your business and your plans for the satellite. They will want to know what areas you want to cover,

The Space Segment

what types of applications will the satellite be used for, the number of transmit and receive sites on the ground, etc. All of these factors will have bearing on the satellite design.

Types of Satellites

Basically, today's satellites are of two design types: Spin-stabilized and Body-stabilized satellites. The first communications satellite successfully launched into geosynchronous orbit, Syncom-3, was a spin-stabilized satellite. Spin stabilization was developed by a team of engineers at Hughes led Dr. Harold Rosen. Spin stabilization enables a satellite to maintain its orbital position by spinning on its own axis at a constant rate, much like a top spinning at a certain speed is able to maintain its position on the ground. Body-stabilized satellites do not rely on the rotation of the satellite to stabilize its position. Instead it has three wheels spinning at very high speeds located inside the box-type body of the satellite to provide stability (See Figure 4.1). The three wheels are guided by a highly precise gyroscope which helps maintain the satellite's orientation. The body of a spin-stabilized satellite can accommodate more antennas and two large deployable solar panels, which means it can provide more services and more power. Body-stabilized satellites are also called "three-axis" satellites.

Most of the satellites from the 60s to the early 80s were of the spin-stabilized design. However the spin-stabilized satellites are limited by the number of antennas and the amount of solar panels you can put on the satellite since the drum-shaped body constantly spins. So you can only put antennas fixed to a position on the ground on the top of the satellite and the solar arrays are wrapped around the body. As the power and other communications requirements by satellite operators became more complex and sophisticated, the body-stabilized satellite design, which can incorporate a greater number and larger-sized

Cut-out image of the Boeing 376 spin-stabilized satellite showing its component parts.
(Source : Boeing Satellite Systems)

57

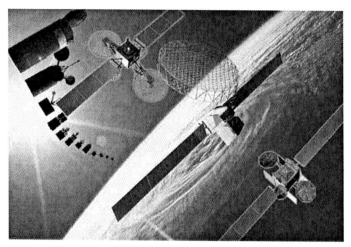

Figure 4.1. Satellites have evolved not only in size and weight but in power and the complexity of applications that it can provide as seen from this graphic from Boeing that shows from far left the first Syncom-3 drum-shaped spin-stabilized satellites to the later Boeing models that are body-stabilized satellites (the box-shaped with large unfolding solar panels). The first Syncom satellite weighed only 140 lbs. and had only one transponder capable of receiving and retransmitting one television program while the latest Boeing 702 model at over 9,000 lbs. can have more than 100 transponders and broadcast over 3,000 television channels. (Image courtesy of Boeing Satellite Systems).

antennas and solar panels has became more or less the standard design for geostationary satellites.

Satellites are also differentiated in terms of size with geostationary satellites weighing at 1,000 kilograms (kg) or about 2,200 pounds (lbs) or more. The latest Boeing 702 satellite model can weigh 4,060 kg or 8,951 lbs at launch and 2,630 kg or 5,798 lbs at the beginning of its orbital life (BOL). The discrepancy in the satellite's weight at launch and at the beginning of its orbital life is the amount of fuel it expends to get into orbit. So the majority of a weight of a satellite is its fuel (much like the rocket that launches a satellite, as we shall see later).

Smaller satellites which usually are launched into mid-earth or low-earth orbits are satellites, weigh generally less than 1000 kg. Further distinctions are those that weigh 100 kg or less are called *microsatellites* while those weighing less than 10 kg are called *nanosatellites* and believe it or not there are *picosatellites* that weigh less than 1 kg (2.2. lbs).

Smaller satellites are more suited to some applications that do not require the sophistication or functionality of the larger models. They are also much cheaper and take lesser time to build, and can therefore be launched in less than a year as opposed to the larger satellites which average about two years to build.

Some countries, for reasons of national pride, just want to get into the satellite bandwagon by having the distinction of launching their own satellite, use smaller satellites launched into low-earth orbit, just so that they can be counted among the relatively few nations who have launched a satellite of their own into space.

Satellite Applications

Satellites in geosynchronous orbit are most suited for broadcasting and high bandwidth data applications because the satellite remains in the same position relative to the earth precluding the need for several antennas on the ground to track the satellite. Therefore a single dish can be pointed to the satellite at all times without having to move to transmit and receive signals to and from the satellite. Satellites in other orbits such as elliptical and low-earth (LEO) and medium-earth (MEO) orbits require tracking by several antennas. The term geostationary orbit should not be confused with the *geosynchronous* orbit, although sometimes both terms are used interchangeably. Geostationary orbits are a type of geosynchronous orbit, where an object moves at the same speed as the earth. However, some satellites using the geosynchronous orbit deviate a little north and south of their orbital path creating an inclined orbit, which technically is not in geostationary orbit. Satellites are intentionally put on inclined orbits to conserve fuel that would otherwise be used to make periodic corrections to its position. Using less fuel prolongs a satellite's service life. The tradeoff is that inclined orbit satellites require tracking by antennas on the ground.

Satellites in geosynchronous orbits also have the advantage of higher elevation than lower-earth orbits. This enables its antennas pointing at the the earth to cover a wider area. Given that satellites in lower-earth orbits have smaller coverage areas, it will require several MEO or LEO satellites to achieve the same coverage by a GEO satellite.

The geostationary arc is dominated by commercial satellite operators that provide mainly telecommunications services (these will be discussed in greater detail in Chapter 6—Satellite Services). However, satellites in different orbits like MEO and LEO perform many other applications and vital services provided by commercial, government and international non-profit organizations such as:

Mobile Satellite Services. Mobile satellite telephones use satellites in lower earth orbit like the Iridium and Globalstar constellation of satellites to provide telephony and data services to remote parts of the world. In lower earth orbit, there is no signal delay unlike in the geostationary orbit where a signal has to travel about 46,000 miles (23,000 to uplink to the satellite and another 23,000 to downlink back to earth) which results in about one-fourth of a second (.22 sec) delay. Because there is no noticeable delay in the signal when using MEO or GEO satellites, they are ideal for mobile telephony which requires seamless, real-time conversations.

Remote sensing. Satellites monitor different parameters on the ground such as levels of radiation, temperature, and geological features of the earth through the use of highly sensitive sensors both in the satellite and on the ground. The data collected by these satellites are vital to environmental and resource planning as well as scientific research.

Imaging/Observation. Similar to remote sensing applications, satellites can take very precise images of the earth's surface that are vital for monitoring and planning purposes. Satellite imagery are now readily available on the internet or can now be purchased from commercial providers. Images taken from satellites can have resolution of as much as 3 feet, which means it can distinguish objects on the ground 3 feet or larger. Most imaging satellites are in lower earth orbit in order to be closer to the earth and be able to take higher resolution images.

Weather/Meteorological. Dedicated weather satellites monitor the earth's atmosphere, weather patterns and tracks storms, hurricanes, forest fires and other natural disasters. These satellites are able to pinpoint the location and time of storms and other weather phenomena. The news media, travel, maritime, aviation and other industries are highly dependent on weather reports culled from satellite sources. The information gathered by weather satellites are also vital in emergency planning and evacuation of threatened populations.

Navigation. Satellites systems like the US military's Global Positioning System (GPS)—a network of 24 satellites in low-earth orbit that can provide the location or coordinates of an object on the ground has proven invaluable as a navigation tool for various marine craft and now available in vehicles and handheld devices. GPS is provide free of charge by the US military, but you'll need a device on the ground to obtain the coordinates of your position in terms of longitude and latitude. User-friendly handheld and vehicle-mounted devices pinpoint your location on a map and provide street-by-street directions on how to get from point A to point B.

The European Union is trying to develop an alternative to the GPS system, with the promise of greater accuracy. The alternative system called "Galileo" will take years to develop and won't probably be operational until 2015, that is, if they surmount many of the financial and technical hurdles that they are facing.

The former Soviet Union built a satellite navigation system called GLONASS that covers most of Europe and parts of Asia, but the system went into disrepair after the break up of the former USSR. Russia has recently partnered with India to revive the system and expand it to have global coverage.

Scientific Research. Satellites can also perform many different scientific research functions mainly in the fields of astronomy, biology and earth sciences. Astronomical observation satellites such as the NASA's Hubble Space Telescope operate in LEO orbit like a ground-based space telescope. Astronomical satellites take advantage of the earth's orbits to provide a clearer view of the heavens unobstructed by light pollution, weather and other phenomena on the ground. Other scientific satellites perform various experiments in space and monitor and observe the earth's geological patterns and phenomena.

In addition, there are dedicated military satellites that perform a variety of all the abovementioned applications for military purposes, as well as specific military communications applications.

All the aforementioned services can be performed in GEO, MEO or LEO orbits, but the different economies of scale dictate that certain orbits are more suited to certain types of applications or services.

Main Components of a Satellite

A satellite has two main components: the "Bus" which contains the power, propulsion, telemetry and control systems and the "Payload" which in telecommunications satellites, contains the communications

61

The Satellite Technology Guide for the 21st Century

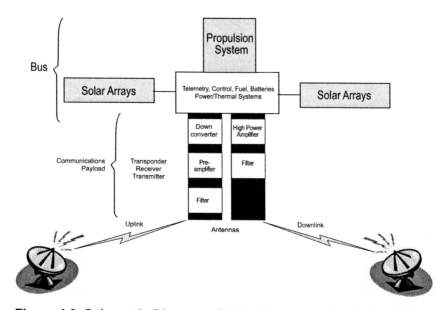

Figure 4.2. Schematic Diagram of Main Components of a Satellite

A typical satellite consists of two main components: the bus and the payload.

systems consisting mainly of the antennas and the transponders of the satellites that receives and re-transmits signals (see Figure 4.2).

Satellite Bus

When you order a satellite, the bus is usually standardized to save time in manufacturing the satellite. As mentioned earlier, the bus consists of the structure (the body that houses the payload components), power systems like the solar panels, batteries, propulsion system and the telemetry, tracking and control systems. Let's discuss each of these systems:

Propulsion system. Satellites have their own propulsion system as it would need to power itself at some point in the launch process to get to the right orbital location and to make occasional corrections to its orbital position. As we will see when we discuss the launch process, a satellite once it separates from the launch vehicle when it is placed in a transfer orbit (an intermediate orbit before it goes into geostationary orbit) will have to power itself through its apogee kick motor to geostationary orbit. Once in orbit—the satellite has thrusters that are powered on occasionally to make adjustments in its position in order to maintain its orbital location. This function of a satellite is called "station keeping," and its purpose is to maintain a satellite in its correct orbital position. The corrections made by using the satellite's thrusters are called 'attitude control.' A satellite in geosynchronous orbit can deviate up to 9/10ths of a degree every year from north to south or east to west of its location due to the gravitational pull of the moon and the sun, so thrusters are needed to make the necessary corrections. A satellite's lifespan is determined by the amount of fuel it has to power these thrusters. Once the fuel runs out, the satellite eventually drifts into space and out of operation.

Power System. A satellite in orbit has to operate 24 hours a day, 365 days a year during its lifespan of 15 years or more. It needs internal power to be able to operate its electronic systems and communications payload. The main source of power of a satellite is the sun, whose power is harnessed by its solar panels. It also has batteries on board to provide power during the 40 percent of the time in the course of its orbital path around the earth, it cannot see the sun when the sun is behind the earth relative to the satellite's position. The batteries are recharged by the excess current generated by the solar panels when there is sunlight. Every year of operation, the solar panels and batteries of a satellite degrades resulting in a gradual loss of power as the satellite ages.

Thermal System. The harsh environment in space results in extreme cold and hot temperatures of between -100°C to +100°C. Thus satellites have thermal control systems to protect the sensitive electronic and mechanical components of the satellite to maintain it in its optimum functioning temperature and to ensure its continuous operation.

Tracking, Telemetry and Control System (TT&C). The TT&C system of satellite is a two-way communication link between the satellite and TT&C on the ground. This allows a ground station to track a satellite's position and control the satellite's propulsion, thermal and other systems. It can also monitor the temperature, electrical voltages and other important parameters of a satellite.

Communications Payload

The communications payload is where you can customize the satellite in terms of the number of transponders, the different frequencies it will be operating in and other electronic systems. A typical satellite communications payload consists of the same equipment chain as in ground systems (as we will see in the next chapter). Like an earth station, satellites have antennas, upconverters and downconverters, multiplexers and high-powered amplifiers to be able to receive and retransmit signals from and to the ground (earth stations do just the reverse—they transmit and receive signals to and from the satellite).

The key component of the communications payload of a satellite is the transponder. A transponder is the revenue generating part of the satellite. Satellite operators make their money from transponder sales or leases and other value-added services they provide on the ground. Satellite transponders are divided into segments of varying bandwidth typically: 27 MHz, 36 MHz, 54 MHz or 72 MHz. Customers can lease a whole transponder of part of it. As we have seen from the previous chapter, each transponder operates in a specific frequency and can have different polarization. Frequency reuse by varying frequency bands and polarization maximizes the bandwidth availability of a satellite—thereby increasing the revenue stream for satellite operators.

Transponder leases vary greatly from satellite to satellite, but average at about US $100,000 per month for C-Band. Satellite operators prefer long-term leases of three years or more, but you can lease a transponder for as little as 15 minutes for transponders reserved for occasional use (or transponders that have not yet been leased long-term and are temporarily earmarked for occasional use in the meantime).

The most basic type of satellite architecture is called a "bent-pipe" which as the term implies basically just takes a signal from the ground and retransmits it back without any processing of the signal. Sometimes a transponder is also referred to as a "repeater" because that is what it essentially does. A satellite's receive antenna receives the signal from the ground in the uplink frequency that it was transmitted, then the satellite's transponder filters this signal and converts it to the downlink frequency and transmits this back to the ground. The workhorse of the transponder is its high power amplifier, which amplifies the weak signal it receives from the ground so that it can be converted to the downlink frequency for retransmission. This is performed by either a Traveling Wave Tube Amplifier (TWTA) or a Solid State Power Amplifier (SSPA).

Some satellite models have sophisticated on-board processing systems in the communications payload that can perform many functions

The Space Segment

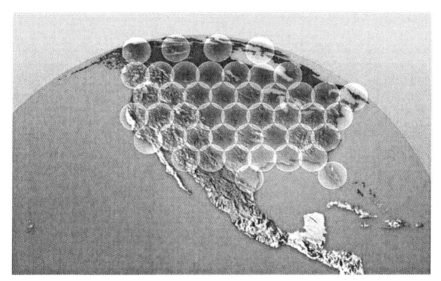

Figure 4.3. Powerful on-board processors of a satellite enable beam switching and concentrated spot beams as can be seen from the above coverage map for the Canadian Anik-2 satellite which provides broadband internet services throughout North America. (Source: Boeing Satellite Systems).

like "cross-strapping" which means it can receive a signal in a certain frequency, i.e. C-Band and retransmit it in another frequency, i.e. Ku-Band. On-board processors switch signals to different "spot beams" and can also have more concentrated beams to specific areas. On-board processors can also improve digital signals through error correction techniques.

The advantage of on-board processing is you can switch a signal's beam to different coverage areas or to a different frequency band, when your business requires a change in focus. For example, at the beginning of your contract you plan to cover say, the Caribbean region in C-Band. However, two years down the road, due to changing business and regulatory conditions you need to switch coverage to Mexico instead and in Ku-Band. A satellite with on-board processing capability can make this happen whereas a "bent-pipe" type of satellite cannot. This also works if you are a satellite operator and need the flexibility to change your beam offerings according to changing market demands.

Even more sophisticated satellites can link with other satellites in different orbital locations. These are called "intersatellite links" and it

The Satellite Technology Guide for the 21st Century

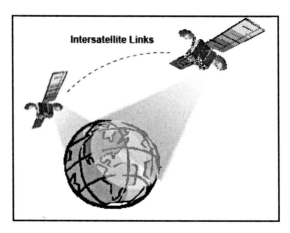

Figure 4.4. Intersatellite links preclude the need for several hops through multiple ground stations and satellite links. This represents significant cost savings as well as lesser time delays in transmission (since the signal has less to travel). There are also lesser links in the chain where errors and signal degradation may occur.

saves on having to do several hops of uplinks and downlinks on the ground (see Figure 4.4). Intersatellite links also lessens the distance that a signal has to travel and therefore lessens the delay inherent in satellite transmissions. Intersatellite links are essentially like "spot beams" except that instead of pointing the satellite's antenna to the ground, it is pointed to other satellites. Intersatellite links can be made not just between GEO satellites but between GEO and other lower earth orbit satellites. However, both satellites should have the capability to either receive or transmit signals from the other satellite in order to communicate.

Satellite transponder capacity can be used by a dedicated single user or shared among multiple users. Different modulation techniques of the signal are used in order to share the bandwidth of a specific transponder. These modulation techniques include Demand Assigned Multiple Access (DAMA), which functions like a telephone switchboard where a user's call can determine temporary access. Another method of access sharing is Time Division Multiple Access (TDMA) where access is divided into time units or Code Division Multiple Access (CDMA) which assigns a code to each user. The most common used method of sharing bandwidth on a satellite is Frequency Division Multiple Access (FDMA) which divides the frequency band for different users.

Leasing Transponder Capacity

Most satellite transponder capacity is leased, or provided as part of a bundled service that includes uplinking and other services. However in situations where a satellite is being marketed prior to launch during the 18-36 month construction phase, the operator may benefit by selling capacity interests outright as part of the overall financing of the project. Here, the buyer "owns" and takes title to the transponder and may depreciate the asset, while the asset is removed from the satellite operators' list of depreciable assets.

Pre-Launch Deals

As in real-estate development projects, good values can be obtained by investing in satellite capacity prior to a spacecraft's launch. This is because the satellite owner has a big incentive to secure commit-

Buying Satellite Capacity – User Considerations

In leasing transponder capacity and services there are a number of factors you should consider beyond price and basic technical performance:

Satellite Neighborhoods

A particular satellite's "Neighborhood" can be a key consideration in the value of satellite capacity. Neighborhood generally refers to the group of TV program networks on a particular satellite's orbital position. If most cable head-ends in a given marketplace have antennas pointed on a particular satellite, then it benefits new cable channels to distribute on that satellite, since it will be easy for cable operators to pick up the signal using their existing downlink systems, and programming is most widely distributed.

The same principle can also apply to communities of international programming and event business networks, where a particular industry, such as automotive may have a large installed base of receive antennas fixed on a particular satellite. The launch of data and video services may be readily deliverable to that industry using the same

The Satellite Technology Guide for the 21st Century

satellite without requiring the downlink sites to install new antennas at each site.

Interference Considerations

Neighboring satellites' degree of interference varies around the world, depending on the location of the ground stations and satellites. Network planners need to work with satellite operators to ensure that the potential for current and future adjacent satellite interference is understood before purchasing a transponder.

Signals on opposite polarizations of a satellite can also cause interference, and the technical impact on a network should be assessed before buying transponder capacity.

Transponder Non-Linearity Characteristics

Another buying consideration is the transponder amplifier linearity, and ability to adjust gain step settings on the transponder, depending on the mix of applications that might be planned for the capacity. A single large video signal occupying the transponder will have less complex performance needs than a service that plans to operate Very Small Aperture Terminal (VSAT) networks, Satellite News Gathering (SNG) and multi-channel video all in a single transponder.

Satellite Health, History, and Reliability

As with any technology and complex system that must operate on solar power 22,300 miles over the equator 24 hours per day, 365 days a year, for 15 years in the extreme hot, cold, electrostatic, environment of space, satellites and transponders can suffer performance impairments and event total system failures. Certain models of spacecraft have proven themselves as reliable, while some subsystems have had patterns of failures. Satellites that have reported degradation or operated on backup systems or non-redundant critical subsystems due to known primary or backup subsystem failures, will be perceived by the market as reduced in value.

Space insurance companies underwrite against certain types of failures or losses of satellites and may report statistics on particular types of satellites and operators' fleets. Publicly traded satellite operators are typically required to disclose impairments to their satellites to inform investors of the firm's assets. Large transponder buyers are usually entitled to reports on satellites health as well from the operator.

Satellite capacity is often marketed based on different levels of service priority. For example, non-preemptible transponders may not be preempted by other transponders, whereas capacity sold as *preemptible* may be offered on the condition that it can be preempted if there is a need to restore other customers whose services are of a higher priority. Generally, the higher the level of protections of transponders, the higher the value and price.

Preemptible means your transponder can be reassigned to other services if needed by the operator, thereby resulting in a loss of service for your network. So, never enter into a preemptible transponder agreement unless you are willing to yield your transponder to the operator on demand.

Intra-Satellite Protection

Intra satellite protection is where a transponder on a satellite can preempt another capacity on the same satellite, in the event the protected capacity suffers a failure. With inter-satellite protection, a whole satellite may be preemptible in order to protect another satellite with a more critical role, or transponders on one satellite may be restorable on another satellite's capacity at the other role in case of a failure.

Fleet Backup Schemes

Some operators can provide a regional "spare satellite" whose function includes the role of restoring services on satellite in the same region that suffers a complete spacecraft failure.

-by Daniel B. Freyer

ted lease revenue as soon as possible after the in-service date. The risk involved here is that you are taking a chance before the satellite is launched. In the event the satellite fails to launch or if the venture goes belly up, you may be stuck with whatever deposit you made on the contract.

Resale & "Brokers"

In some cases, it is beneficial to purchase satellite services from resellers or brokers. Just as real estate brokers market properties in exchange for a sales commission, satellite resellers and brokers help market unused capacity for a commission. Companies like Vista Satellite Communications and Space Connection are leading brokers of satellite capacity in the U.S. and hold inventories of transponders on multiple satellites through their own leases or companies which they represent. Online brokers like the London Satellite Exchange also provide brokering services (www.satellite-exchange.com). In the same manner, brokers can help identify specific space across multiple satellite operators, identify space that's not otherwise for sale on the open market for a buyer that wishes to be "discrete," and advise capacity buyers on market rates across different satellite fleets.

Full-Time vs. Partial Transponder Services

Transponder capacity is typically sold in either full- or partial transponder increments. Most networks require 24 hour dedicated use of satellite transponder capacity. The traditional full transponder is a physical channel unit, having a specific usable segment of bandwidth, typically 27 MHz, 36 MHz, 54 MHz or 72 MHz. The different segments may also have access to different amplifier options. When leased or purchased, the user may have the ability to use the full power (EIRP) and bandwidth (MHz) resource of the transponder.

Pricing for capacity varies according to supply and demand in the marketplace. A recent rate in 2006, for one U.S. operator, if booked online for 3.0 MHz of domestic U.S. C-band was US $15.00 for 15 minutes, compared to US $63.00 for 15 minutes on a high-powered Ku-Band transponder over Asia for the same 3.0 MHz. This pricing cover only the cost of the transponder lease, or what is called the "space segment" of the uplink chain. You will still have to pay uplinking charges from a teleport, which forms part of the "ground segment" of the chain.

A "partial transponder" consists of lease bandwidth and power resource on a transponder which may be shared with other users. Essen-

The Space Segment

tially the customer is buying the ability to operate a pre-determined carrier (or set of possible) carrier options into the leased bandwidth.

Pricing for partial transponder capacity is typically based on the number of MegaHertz leased per month, factoring in a power usage (aggregate leased carriers EIRP) within the bandwidth (MHz) used on the satellite. For instance, $4,000 per MHz/month might be a C-band rate in some markets. Another similar way of pricing for partial transponder capacity is based on the greater percentage of power or bandwidth used. For example if a TV signal required 10% of the transponder's calculated power and 15% of the bandwidth, then the price would be based on 15% times a standard price per percentage per month.

For applications like VSAT, data, digital business TV, single-channel video and radio distribution and contribution networks–most of which use carriers of less than 5 Mbps, only a partial transponder service is typically required.

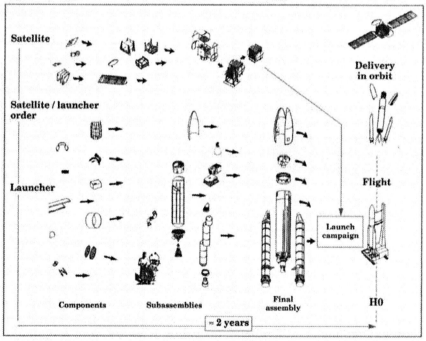

Figure 4.5. The Satellite Launch Cycle. The assembly and construction of various components required to launch a satellite involves many sub-contractors and can take up to two years. (Image courtesy of Arianespace).

The Satellite Technology Guide for the 21st Century

Full-Time vs. Occasional and Recurring

Most networks require dedicated "permanent" or continued use of a transponder capacity. For example, because HBO requires transponders 24 hours a day, 365 days per year to deliver its channels to cable TV affiliates, its parent company Time Warner has leased many of its transponders on long-term contracts of over 10 years.

However, many applications, particularly in television only require occasional, *ad hoc*, or part-time use of capacity, and for these uses, satellite operators and service providers sell capacity by time fractions of hours, and even in some cases as little as a few minutes. Partial transponder capacity, for instance, can be purchased from satellite operators in C- and Ku-Band in increments of 3.5 MHz to 36 MHz. Or it can also be purchased in increments of 15 minutes to thousands of hours per year.

Launch Services

Contracting a launch service provider, should be done in tandem with purchasing a satellite from a manufacturer. Satellite launch service providers do not have on-hand an inventory of rockets needed to launch a satellite.

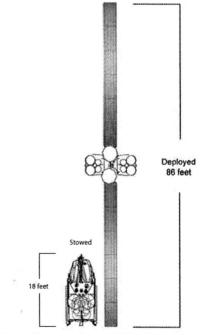

Figure 4.6. Comparison of the height of a Boeing 702 satellite when stowed and upon deployment.

Rocket launchers are very complex and expensive to build, so they are assembled only upon signing of a launch contract. Figure 4.5 shows the process of launching a satellite which can take 2-3 years, if done in conjunction with the construction of a satellite.

The rockets that launch satellites have to be built and assembled once a satellite launch service is ordered. Ideally the launch service contract should be signed at about the same time as the satellite manufacturing contract. This will ensure that when your satellite is completed it won't have to wait too long till it is launched. Launch service providers have a backlog of launches scheduled and getting a launch date requires advance planning.

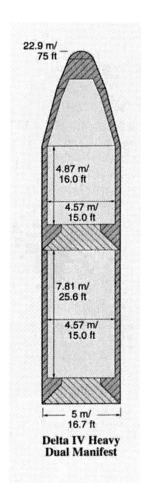

Once a satellite is completed and has undergone rigorous testing in the manufacturing facility, it is placed in its stowed position and shipped to the satellite launch facility. The satellite is in the "stowed" position to save space so that it can fit into the launcher's fairing—the cone shape tip of the launch vehicle where the payload is stored. In launch terminology, the "payload" is the cargo it has to carry to space, which in this case is the satellite spacecraft. The satellite will "deploy" its solar panels and antennas only after it has been successfully launched into orbit.

To be able to escape the earth's gravitational pull and be placed into geostationary orbit, a satellite has to be launched at a speed of 17,500 miles per hour to a height of about 22,000 miles above the earth. This requires a tremendous amount of fuel. The majority of a launcher's mass is its fuel. Only 1-2 percent of the launcher's mass is its payload. That means that it requires a 200-ton rocket to launch a 2-ton satellite.

Figure 4.7. Today's rockets are very powerful and can carry multiple satellites in a single launch vehicle. The Boeing Delta IV rocket for example, can carry up to three geostationary satellites or several smaller satellites. (Source: Boeing Satellite Systems).

The Satellite Technology Guide for the 21st Century

What enables launchers to travel vast distances is the use of multi-stage rockets. This technique was initially conceived in the late 19th century by a Russian scientist named Konstantin Tsiolkovsky. Multi-stage rockets enabled the first forays into space in the 60s that lead eventually to the landing of the first human on the moon in 1969. A multi-stage rocket is essentially several expendable rocket modules that are integrated into one vehicle. Each rocket module can function independently and is assigned a stage in the mission. Once a rocket's stage is over (and all its fuel is expended) it is jettisoned into space. This conserves the mass of the vehicle, thereby requiring less and less power as the mission progresses. After all the rocket modules are jettisoned, the satellite is eventually separated from the launch vehicle and propels itself under its own power to its final orbital position.

The most crucial stage is the liftoff from the ground to about 300 miles to low earth-orbit. The first stage of the rocket usually includes booster rockets strapped to the side of the main rocket. The launch vehicle goes to low-earth orbit prior to the higher geosynchronous orbit. This method of launching into higher orbit by going through lower intermediate transfer orbit is the most efficient way of launching using a minimum amount of energy or fuel. Once in the lower transfer orbit, fuel is expended only at the apogee or perigee of the orbit (the highest and lowest point of the orbit) in order to transfer to the higher orbit. This maneuver was conceived in 1925 by a German engineer named Walter Hohmann and is called "Hohmann's transfer."

The initial launch sequence of a satellite lasts about an hour and half. If you have not seen a rocket launching into space—it is truly one of the most spectacular sights on earth and I highly recommend it. However, for the owners of satellites and their customers as well as the launch service providers, that crucial first hour in a satellite launch can be the most gut-wrenching experiences they will ever have. In a span of an hour or less, hundreds of millions of dollars in investment as well as years of labor can literally go up in smoke. Fortunately, launch technology has continually improved over the years and the great majority of launches usually end up in success. The launch industry's rate of success in the last few years averages about 94 percent.

Figure 4.8 shows the main sequence of events in a satellite launch. The launcher lifts off from the launch site and upon exiting the earth's atmosphere, its fairing—the cone shape tip of the launcher where the satellite is stored, is jettisoned. The launcher goes into a transfer orbit, which is a preliminary orbit where the satellite can be fired into the

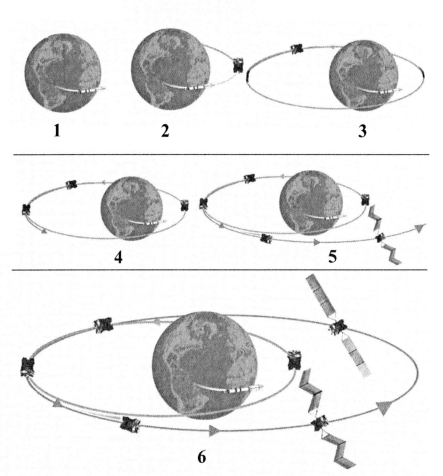

Figure 4.8. Geostationary Satellite Launch Sequence. After (1) liftoff, the different stages of the (2) launch vehicle are jettisoned followed by the (3) separation of the satellite from the launch vehicle when it reaches transfer orbit. At the (4) apogee or highest point of the transfer orbit, the satellite's apogee kick motor is fired to propel it into geostationary orbit where (5) its antennas and solar panels are deployed and then satellite (6) undergoes in-orbit testing before it starts operation.

geostationary orbit. The satellite is then separated from the launch vehicle and goes into the transfer orbit. At the apogee or the highest point of the orbit, the satellite's apogee kick motor is fired to take the satellite to geostationary orbit where it then deploys its antennas and solar panels.

In-orbit testing of the satellite follows its successful launch into geostationary orbit. This in-orbit testing process can take from a month to several months before a satellite can be operational and begin service.

Satellite Insurance

Finally, since so many things can go wrong with a satellite from launch to its long operational life, it's imperative to insure the satellite. The satellite can be insured for launch failure and even in-orbit anomalies or malfunctions. Satellite insurance costs can be very hefty, but there's no going around it. Premiums can run into the tens of millions of dollars per year. The satellite manufacturers and launch service providers do not assume any of the cost of insurance, although there may be penalties built-in the contract for delays or equipment malfunctions. The responsibility for insuring the satellite lies with the satellite operator.

As a major investment asset, satellites simply have to have some form of insurance. For GEO satellites that cost hundreds of millions of dollars and several years to build, insurance is an essential cost that has to be factored in a satellite's business plan.

CHAPTER 5

The Ground Segment

Dr. Joseph Pelton, one of the foremost scholars of the satellite industry, perhaps gave the best analogy for a satellite communication network. He said that in satellite communications "it takes at least three to tango." Indeed, the most basic satellite link requires a minimum of three elements to complete the loop: a sender on the ground uplinking a signal—a satellite in space receiving and retransmitting the signal back to the ground—and a receiver downlinking the signal in the ground. In the previous chapter, we have discussed the "space segment" part of a satellite network. In this chapter, we will discuss the "ground segment" or the components on the ground that completes a satellite network.

As we have seen, all satellite signals emanate from the ground using transmission equipment and antennas to send a signal to a satellite in space. The same antennas can also receive signals on the ground from the satellite. The generic term for satellite transmission and receive equipment on the ground are called "earth stations." Sometimes terminologies can be confusing as "earth stations" are used interchangeably in satellite parlance to mean a "teleport" or just the antenna or "dish" or the "terminal" (the transmit and receive equipment). So, for example, a Very Small Aperture Terminal (VSAT) equipment consisting of a small dish and receive equipment used in many businesses to send and receive data and video over satellites are interchangeably referred to as "earth stations" or as "terminals." A teleport, which is a ground facility, is sometimes referred to as an "earth station" when what it really is a collection of many earth stations. To avoid confusion, an earth station is a single unit of receive or transmit (or both) equipment consisting of an antenna and transmission and/or reception equipment. So one VSAT ter-

The Satellite Technology Guide for the 21st Century

minal is an earth station, while a teleport is an earth station providing multiple uplinking and downlinking of signals from several satellites.

I prefer the term "user terminals" (UTs) instead of what some would refer to as "earth stations" or simply terminals. UTs include VSATs, handheld mobile satellite telephone sets, portable GPS navigation sets, satellite radio receivers and the like. All of these sets are able to receive and transmit to satellites, despite their relative difference in size.

Earth stations can be of two types: *fixed* or *mobile*. Fixed earth stations are permanently installed in a specific area, like in teleports, network centers or in broadcast facilities, while mobile earth stations can be moved from one location to another. Mobile earth stations are now

Advances in mobile satellite receive and transmit equipment enable these broadcast journalists to maintain voice and e-mail contact with their studios and provide broadcast-quality videos using only a small flat panel antenna and a laptop provided by Inmarsat's BGAN service (photo courtesy of Stratos Global).

so portable that they can fit in a small briefcase and can be used for many different applications such as Satellite News Gathering and Emergency and Disaster Communications. Mobile earth stations can be mounted on any vehicle or vessel and be operated even while they are moving.

To function, an earth station must have a clear line of sight to the satellite so that the transmission and reception of a signal are unobstructed and unimpeded. Teleports are usually located in places where there is little chance of interference from other terrestrial sources of radio signals and microwaves.

Earth Station Architecture

The basic configuration of a typical earth station like a teleport consists of five main subsystems: the antenna subsystem; the RF equipment subsystem; the ground communications subsystem; the terrestrial interface subsystem and the support systems (see Figure 5.1). Let's discuss each of these subsystems:

The Antenna subsystem. The satellite antenna or dish is one of the most recognizable icons representing a satellite communication network. They can range in size from a few inches in diameter for Ka- and Ku-Band dishes to over 100 feet (30 meters) in diameter for C-Band dishes. The size of the antenna is dependent on the power of the satellites it is looking at and frequency band it operates in. There are many types of antennas, but the most commonly used satellite antenna is the parabolic type. An antenna can receive or transmit signals from a satellite. To do this it has to be pointed directly and precisely at a satellite and it must have enough power to do so. Usually, each dish can only "look" or communicate with one satellite, due to various factors such as the look angle to the satellite, the coverage of the satellite, and the power of the signal to and from the satellite. There are special dishes that are able to look at many satellites. Mesa, Arizona-based ATCi, for example, manufactures the Simulsat brand of antenna that can look at up to 35 satellites within a 70 degree range of the geostationary arc. This type of antenna is pricier than most antennas, but saves a lot of valuable real estate in the antenna farm and provides greater flexibility for teleports to provide services to various satellites.

The antenna is the first link in the ground segment side of the satellite network when transmitting and receiving signals to and from the satellite. To receive a signal from a satellite, an antenna reflects the weak signal on its dish to a feedhorn which then goes through a Low Noise Amplifier (LNA) to amplify the weak signals. This signal is then

The Satellite Technology Guide for the 21st Century

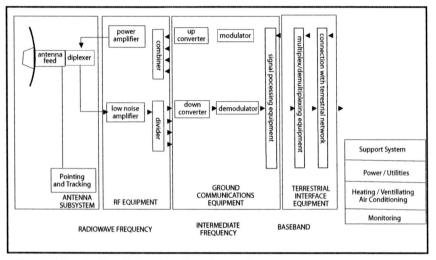

Fig. 5.1 Basic Earth Station Configuration

demodulated or frequency converted by the RF equipment subsystem. Receiving a signal from a satellite is called "downlinking" while transmitting a signal is called "uplinking." To uplink a signal the antenna's feedhorn takes the signal amplified by the High Power Amplifiers (HPA) from the RF equipment subsystem and is reflected into dish to the satellite. Because the downlink and uplink operates in different frequencies, an antenna can receive and transmit signals at the same time. Each antenna can only receive or transmit in one frequency band i.e. C-Band, Ku-Band, etc. However, there are antennas that can receive and transmit in both Ku- and C-Band or other frequencies. These are called "dual-band" or even "triple-band" antennas. Their ability to do so is determined by the amount of ancillary equipment they have.

Of course not all antennas can receive and transmit at the same time. Some antennas, like those used in consumer Direct-to-Home (DTH) satellite services, can only receive signals and are called TVRO (Television Receive Only) antennas. Very Small Aperture Terminals (VSATs), which are mainly used for businesses, are small antennas that can both receive and transmit signals (VSAT systems will be discussed in greater detail in Chapter 7).

The EIRP (Effective Isotrophic Radiated Power) of a transmit antenna on the ground is an important parameter as that of the signal of a satellite (which is downlinked from the satellite's antenna). So is the

The Ground Segment

Gain/ Temperature (G/T) of a satellite's signal, discussed in the previous chapter, which is important to the reception equipment of a ground antenna. Satellite operators like Intelsat have stringent standards for ground antennas based on the EIRP and G/T, among other parameters. Antennas that conform to operators' standards are considered "type approved." Antenna manufacturers often get type approval from various satellite operators which can act like a "seal of approval" for their products.

RF equipment subsystem. The RF or Radio Frequency subsystem consists of high power amplifiers for the transmit equipment and low noise amplifiers for the receive equipment. Amplifiers, as the term implies, basically amplifies or increases the power of a signal. Before it is transmitted to the satellite a signal has to be amplified as it has to traverse over 22,000 miles to reach the satellite. Conversely a signal from a satellite is so weak by the time it reaches the ground that it has to be amplified again so that it can be converted into an acceptable signal that can be received by ground equipment. To give you an idea of how weak the signals are from a satellite, they are typically one microwatt in power (or one-millionth of a watt). There are three type of high power amplifiers or HPAs used by earth stations: Klystron Power Amplifiers (KPAs); Travelling Wave Tube Amplifiers (TWTAs) and Solid State Power Amplifiers (SSPAs). All of these perform the same function—to amplify the signal before it is transmitted to the satellite.

The low noise amplifier or LNA of an antenna's receiver basically takes a very weak signal, eliminates the "noise" or unwanted signals and converts it to the intermediate frequency or IF, usually at 70 MHz. IF signals are much lower frequencies than satellite frequencies and travel through cables with much less attenuation or degradation. It is also much easier and cheaper to design electronic circuits to operate at these lower frequencies, rather than the very high frequencies of satellite transmissions. Thus, IF frequencies are the preferred mode of transmission in earth stations.

Ground communications subsystem. The ground communications subsystem takes the baseband signal from terrestrial sources and converts the signal's frequency to an intermediate frequency range. Then depending on the service requirement, it either encrypts the signal or compresses it using a modulator or compression equipment and then converts it to the satellite's frequency using an upconverter before it is amplified by the High Power Amplifier. In the receive part of the system,

The Satellite Technology Guide for the 21st Century

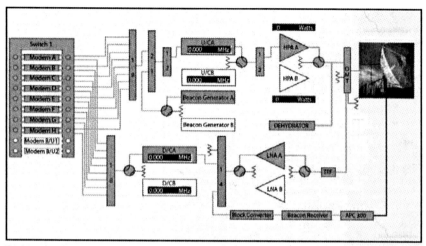

This graphical user interface (GUI) of an earth station monitoring system shows the schematic diagram of various equipment in an earth station. (image courtesy of IDB Systems)

the signal from the satellite is downconverted to IF frequency then demodulated before it is sent to the terrestrial interface equipment for delivery to its final destination.

Terrestrial interface subsystem. The terrestrial interface equipment part of a teleport contains the connections to the public switch network (the telecommunications or telephone network) the fiber network or by microwave or even via IP networks (the internet). Teleports can take signals from remote locations such as broadcast studios, corporate offices and even other teleports and uplink them to a satellite or receive a satellite signal and retransmit through the terrestrial network.

Various signals from various terrestrial sources are routed to the different ground communications equipment and monitors in a teleport using switchers and routers. These signals are then multiplexed (or combined) and demultiplexed using multiplexers and demultiplexers (also called "mux" or "demux" for short). For example, a television signal can be multiplexed to combine the video signal with the audio and demultiplexed at the receive end.

The terrestrial interface system of a teleport is now almost as important as the satellite transmission and reception part of the system with the growing popularity of *hybrid* networks in the provision of satellite services. Hybrid networks use both satellite and terrestrial means of

The Ground Segment

delivery and require interconnection between different networks i.e. satellite and terrestrial.

Support systems. All teleports require redundant backup systems to ensure continuous, uninterrupted operation. Most services require 24/7 operation thus the need for round the clock staffing. This requires Uninterrupted Power Systems (UPS) and backup generators in case of power outages. Teleports, like any facility, require utilities such as telephone, internet access, heating, ventilation and air conditioning systems.

A typical master control room in a teleport includes monitors projected on a screen to enable master control operators to view the quality of the signals being transmitted and received by the facility.
(photo by author taken at the then GlobeCast teleport in Madrid, Spain)

The nerve center of a teleport is the master control room where all the monitoring equipment is located. These are usually manned by operators whose main job is to monitor the signals from various services and other key indicators to ensure the quality of the service.

Teleports

The World Teleport Association (WTA) estimates the annual revenues of teleports at US $15 billion in 2006 or almost 15% of the world's

The Satellite Technology Guide for the 21st Century

Table 5.1 Top 20 Teleports (2006)
(Source: World Teleport Association)

1. **Intelsat** (USA) Through a global fleet of 50+ satellites and teleports based in the US, Europe and Asia, Intelsat offers communications services and managed solutions to customers in approximately 200 countries and territories.

2. **SES Global** (Luxembourg) SES Global is a network of satellite operators reaching 95% of the world's population, operating mainly through SES ASTRA in the EMEA region and SES AMERICOM in the Americas, which together own 28 satellites and multiple teleports in the US and Europe.

3. **Eutelsat** (France) Eutelsat provides capacity on 23 satellites that offer television and radio broadcasting for consumers, professional video broadcasts, corporate networks, Internet services and mobile communications, and operates teleports and uplink facilities in more than a dozen European countries.

4. **GlobeCast** (France) A subsidiary of France Telecom, GlobeCast is a content management and delivery company operating a network of teleports and fiber to manage and transport 10 million hours of video and other rich media yearly.

5. **Telesat** (Canada) Telesat operates a fleet of satellites and a network of teleports in Canada and Latin America for the provision of broadcast distribution and telecommunications services.

6. **Stratos Global** (USA) Stratos offers customers operating in remote locations a variety of wireless, IP, data, and voice solutions, and serves an array of diverse markets including government, military, media, aeronautical, industrial, recreational and maritime users anywhere in the world.

7. **JSAT Corporation** (Japan) JSAT is a leading satellite operator in the Asia-Pacific region, with a fleet of 9 satellites in 8 orbital slots. JSAT was Japan's first private-sector satellite communications operator.

8. **Shin Satellite** (Thailand) Shin Satellite operates five satellites and provides transponder leasing, Internet services and telephone services from its Thaicom Satellite Station near Bangkok.

9. **Space Communications Corp.** (Japan) SCC operates four SUPERBIRD satellites and Japan's largest teleport, SCC Teleport Center in Tokyo, providing satellite broadcasting, telecommunications, DirecPC and DAMA network services.

10. **Loral Skynet** (USA) With its fleet of satellites and its established hybrid VSAT/fiber global network infrastructure, Skynet offers a unique source for

all broadcast, data network, Internet access, and IP needs. (Editor's Note: In 2007, Loral Skynet merged with Telesat Canada.)

11. **Arqiva Satellite Media Solutions** (UK) With teleports and other facilities throughout the greater London area, Arqiva provides permanent and occasional broadcast services as well as IP, voice, data and digital media networks.

12. **Arabsat** (Saudi Arabia) Backed by the 22 member countries of the League of the Arab States, Arabsat offers the Arab World satellite-based communications services including DTH, telephony, Internet and VSAT services through 4 satellites as well as teleports in Saudi Arabia and Tunisia.

13. **Globecomm Systems** (USA) Globecomm integrates satellite into network applications in order to provide reliable, high-quality connection to the edge of the network, broadcast one-to-many, and support bandwidth-hungry applications for media & entertainment, telecom, enterprise and government markets.

14. **Hispasat** (Spain) Hispasat operates seven satellites serving Europe, the Americas and North Africa and providing broadband access to the Internet, interactive and multimedia services, and high quality video conferencing.

15. **Asiasat** (China) AsiaSat operates three in-orbit satellites and Hong Kong-based teleports providing access to more than 50 countries and regions and two-thirds of the world's population.

16. **CapRock Communications** (USA) CapRock delivers world-class satellite communications to the world's harshest and most remote locations. With teleports in the US, Europe and Asia, CapRock provides services that enable its clients to communicate in real-time virtually anywhere in the world.

17. **ND SatCom AG** (Germany) ND SatCom is a global supplier of satellite based broadband VSAT, broadcast and military communication network solutions and services, and operates teleports in Europe and the Middle East.

18. **Telenor Satellite Broadcasting** (Norway) TSB provides satellite communications to the Nordics, Europe and the Middle East via its own and other satellites and a network of teleports in Europe and the United States.

19. **Vyvx** (USA) With more than 16 years of experience with mission-critical content, Vyvx (a unit of Level 3 Communications) provides fiber-optic and satellite video delivery solutions through multiple transmission formats and platforms, including three US-based teleports.

20. **Ascent Media Network Services** (USA) With teleports in the US and Asia, Ascent Media Network Services, part of Ascent Media Group, provides solutions for the management and distribution of content to major motion picture studios, independent producers, broadcast networks, cable channels and other companies.

The Satellite Technology Guide for the 21st Century

Top 20 Independent Teleports* (2006)

1. GlobeCast (France)
2. Stratos Global (USA)
3. Arqiva Satellite Media Solutions (UK)
4. Globecomm Systems (USA)
5. CapRock Communications (USA)
6. ND SatCom AG (Germany)
7. Vyvx (USA)
8. Ascent Media Network Services (USA)
9. RR Sat Global Communications (Israel)
10. Uplit/Petrocom (USA)
11. Globalsat (USA)
12. TIBA (Argentina)
13. Samacom (United Arab Emirates)
14. Essel Shyam Communications (India)
15. Skyport International (USA)
16. ATCi (USA)
17. Datasat Communications (UK)
18. Emerging Markets Communications (USA)
19. Telecommunications Systems (USA)
20. Newcom International (USA)

*Independent teleports are those not affiliated with satellite operators.

Source: World Teleport Association.

total satellite communications revenues. Teleports do not just provide uplink and downlink services, but a host of other value-added services. Teleports provide a wide variety of services that a lot of teleports position themselves as a "one-stop shop" facility where they provide under one roof all the services a client may need to produce and operate a broadcast channel or data network.

The Ground Segment

The most common image of teleports is the antenna farm pictured above. Each antenna is pointed to one satellite and can either receive or transmit signals or both. (photo taken by the author with permission at the then-GlobeCast teleport in Madrid, Spain).

Among the services provided by teleports, according to the WTA, include the following:

• **Origination and distribution of TV channels** to network affiliates, cable Multiple System Operators (MSOs) and Direct-to-Home (DTH) satellites in both real time and via store-and-forward;

• **Television and radio program** production, post-production and hosting;

• **Narrowcast DTH services** provided on a turnkey basis (program origination, subscriber sales, installation, billing and customer service) to ethnic and other narrow TV markets;

• **Hybrid enterprise and government network integration and connectivity** to link remote facilities into the enterprise network for oil & gas exploration and production, mining, aid and disaster relief efforts, and other applications;

• **Distribution of high-value data and video feeds** to closed distribution networks for financial services, hotels, retail stores and malls, bars and restaurants, and other specialty markets;

• **Digital content development, management and distribution** through multiple private and public networks;

The Satellite Technology Guide for the 21st Century

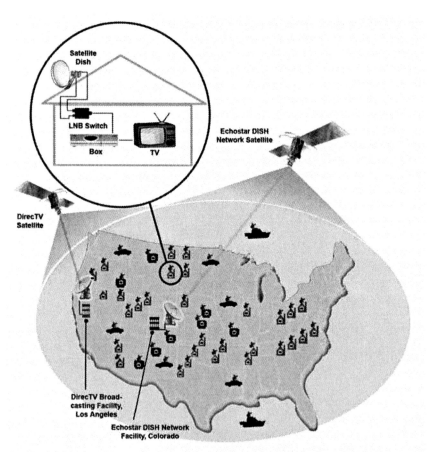

Fig. 5.2 Point-to-Multipoint Distribution of Satellite Signals.
The best illustration of point-to-multipoint application of satellite technology is the popular Direct-to-Home (DTH) Broadcasting Service—where the signal emanating from a broadcast center is uplinked to a satellite and can be received by anyone in a wide coverage area with the right equipment and proper authorization.

• **Distance education and training networks** for enterprises and government;

• **Internet and VoIP backbone services** to Internet Service Providers (ISPs), enterprises and government;

• **Mobile Telephony backhaul** in regions underserved by fiber.

In addition to the aforementioned, teleports provide other services including call centers and customer service support operations; production studios and editing suites; translation and subtitling services; leasing of office space to clients; satellite monitoring and control or Telemetry, Tracking and Control (TT& C), and many others. In addition, teleports provide connectivity links to terrestrial networks. Teleports can virtually provide a turnkey service whereby potential clients can basically meet all their requirements under one roof.

Teleports in the U.S. began in the 70s as small mom and pop operations, some with just a single dish providing uplink and downlink services. However, as competition among satellite operators intensified, consolidation among teleports led to many mergers which created large companies specializing in providing teleport services from multiple locations. The added complexity of requirements from customers which would require heavy investments in terrestrial interface and other equipment also made it uneconomical for the small teleports to survive.

In the new millennium, an emerging trend is that satellite operators have also become teleport operators. Operators which previously only had only satellite monitoring and control facilities on the ground, began purchasing extensive ground facilities in order to be able to provide more value-added services and bundle their space segment with ground segment services. Thus many teleports that flourished in the 80s and 90s have now become part of satellite operators such as Intelsat, SES Global, and others. This is why the WTA in its annual ranking of teleports maintains two lists—the Top 20 teleports which include all teleport operators including the major satellite operators and another list for independent teleports that are not affiliated with any satellite operator (see Table 5.1). Of the top 10 teleports in the list, eight are affiliated with satellite operators.

The main argument against satellite operators that are also teleport service providers is that they tend to provide solutions that are bundled with the provision of the space segment from their own satellites. This can go either way as one might actually benefit cost-wise from a bundling of the ground and the space segment from a single operator. Teleports claim to be "space segment agnostic" and are committed to providing the best solution regardless of satellite space segment. However, if a teleport is affiliated with a satellite operator, it's hard to imagine them going to their competitors since satellite operators rarely have good relationships with their competitors. The thing to remember is to always shop around and get competitive quotes. Sometimes the solution provided by a teleport affiliated with a satellite operator can be the best and the most economical, but always compare different solutions provided by various service providers.

Advantages of Satellites

Now that you already have a good idea of how satellite technology and the main segments of the industry works, it's important to be aware that satellite technology is just one of many different ways of delivering video, voice, data, IP and other telecommunications services. Satellites are facing competition from terrestrial technology such as fiber, cable, copper, microwave, the internet and even new emerging terrestrial wireless technologies such as Wi-Fi and Wi-Max.

The good news is the continued popularity of hybrid solutions that will always include some satellite component to the equation, especially on long-haul or services that require wide coverage. However, in a head-to-head competition with other technologies, satellites have a distinct advantage—it's the best solution for *point-to-multipoint* distribution, such as broadcasting a signal from a single source to many receivers (see Fig. 5.2).

A single satellite can cover up to 40 percent of the world's land area. Three satellites can cover almost the whole world. Satellites are therefore the most cost-effective way to reach a very wide coverage area. In contrast, to wire and connect every home and office in a comparable coverage area of a typical satellite would be prohibitively expensive, especially in the more remote and less densely populated rural areas.

To highlight some of the distinct advantages of satellites:

• Ubiquity of Coverage—Satellites cover everything in its footprint, much like sunlight on a clear day covers the entire surface of the part of the earth in its footprint. Terrestrial systems have to wire every house and building in order to provide ubiquitous coverage. Whereas, anyone with the right equipment and proper authorization can receive satellite signals within the satellite's coverage footprint. It can be more economically feasible, for example, to reach multiple sites using satellite technology than traditional terrestrial solutions.

- Shorter Time to Market—Satellite solutions can be set-up relatively faster than building new infrastructure on the ground. It often also involves less capital outlay and the shorter rollout means revenues can be realized earlier.

- Flexibility and Scalability—Satellite networks are very flexible and can be reconfigured to meet changing demands and requirements. They are also very scalable because you can install temporary networks while testing new markets and expand later on according to the demand.

- Maritime coverage—although some terrestrial networks reach up to a certain point near shorelines, no other technology is able to provide maritime coverage to ships at sea. Since three-fourths of the earth is covered in water, this is a distinct advantage and a unique market for satellite services.

- "Last Mile" Solution—Satellite provides connectivity solutions for that "last mile" directly into individual homes or offices at a fraction of the cost of terrestrial solutions.

- Transcends National Boundaries—Satellite coverage does not distinguish between national boundaries and can provide coverage across countries and regions.

- Coverage during Emergencies and Disasters—As satellite technology is space-based, it is often unaffected by emergencies and disasters on the ground. Coupled with the use of mobile satellite networks on the ground, satellites can provide reliable communications during natural and man-made disasters such as earthquakes and terrorist attacks when terrestrial networks and powerlines are down.

Disadvantages of Satellites

Of course, satellite technology has its disadvantages, too. To name a few:

- High Entry-Level Cost—Starting a satellite network from scratch can be a very expensive proposition. Not to mention there could be regulatory barriers to surmount as well.

The Satellite Technology Guide for the 21st Century

- Latency—One major technical disadvantage of satellites is and inherent delay in transmission because it has to travel 46,000 miles into space and back to the ground. Even at the speed of light, this results in about a fourth of a second (.22 sec) delay in transmission. If the signal has to go through several satellite hops then this delay is magnified. While there are ways to compensate for this delay, it makes some applications that require real-time transmission and feedback such a voice communications not ideal for satellite.

- Prohibitive Replacement/Recovery Cost—As we have seen from the previous chapter on the space segment, satellites are prone to accidents at launch or during its service life in space. Even if insured for the full amount, a satellite lost in space is hard to replace. It can take a couple of years to build a replacement, setting back the timetable for rollout of services. Also, as mentioned in the previous chapter, if a satellite malfunctions in space during its service life the possibility of repair is remote. That's why satellites are built on the premise that it should be able to function for 10-15 years maintenance-free.

Whatever drawbacks satellite technology might have, the advantages far outweigh the disadvantages. Satellites play a very important role in society and have become an indispensable part of modern life.

In the final analysis, there are certain applications such as broadcasting that are ideal for satellite technology, but not all applications are suited for satellite technology. As mentioned earlier, hybrid solutions are becoming the norm where one can get the best advantage from each of the unique strengths of various media and the consensus seems to be that satellites will always have a role to play in any future network architecture.

CHAPTER 6

Satellite Services
By Daniel B. Freyer

Satellites provide a variety of services for communications between fixed transportable, or mobile ground terminals. Service offerings can vary depending on the type of satellite, and its communications payload, or type of transponders on board. However, the most important service provided by satellites is wide area broadcast distribution of information on a point-to-multipoint basis (or, one source to many points of distribution).

Broadcasting via satellite is the "killer" application. Satellite broadcasting continues to have a profound effect on societies around the world as a key element of mass media technology. The number of satellite-to-home, and satellite-to-cable delivered TV channels is estimated to have grown to nearly 12,000 channels over the past 30 years. In addition, the Satellite Industry Association estimated satellite services as a US $52.8 Billion market in 2005—which accounts for nearly two-thirds of the total revenues of the entire satellite industry.[1]

Analog vs. Digital

Prior to the implementation of digital transmission systems, most satellite signals and networks used analog signal modulation for voice, data, and video. Today, virtually all telecommunications traffic on satellites use digital signal modulation.

Although new networks built today use digital signal transmission, there remains a rapidly declining number of networks in the field with analog equipment. Of the more than 25,000 satellite television, radio, and data signals counted on SatcoDX' website in 2006, only 191 were still in analog, or less than 1% of the total signals on satellites.[2]

The Satellite Technology Guide for the 21st Century

In the US, all television signals are mandated by law to be broadcasted digitally by February 2009. The transition from analog to digital broadcasting in the US presents a unique opportunity for the satellite industry to increase its share even further in the digital broadcast transmission market.

Fixed Satellite Services

Fixed Satellite Services (FSS) refers to the authorized service permitted under the International Telecommunications Union (ITU-T) agreements within specific satellite communications radio frequencies for earth-to-space and space-to-earth communications. FSS services uses the following frequencies: C-Band (5.920-6.425 GHz uplink, 3.700-4.200 GHz downlink), Ku-Band (14.000-14.500 GHz uplink, 11.700-12.500 GHz downlink), Ka-Band (27.5 GHz-31.0 GHz uplink, 18.3-20.2 GHz downlink) and X-Band (7.900-8.395 GHz uplink, 7.250-7.745 GHz downlink).

Direct-to-Home (DTH) or Direct Broadcasting Satellite (DBS) services are the priority application planned for the Broadcast Satellite Services (BSS) band, although many such applications are provided the world over using transponders in the conventional FSS Ku-Band.

X-Band, at a higher frequency than Ku or Ka, is traditionally reserved for military satellite communications in the US. However, a commercial venture called Xtar—a joint-venture between Loral and the Spanish operator, Hispasat, provides commercial X-Band capacity for various applications.

Mobile Satellite Services (MSS) are authorized in other radio frequency bands such as L-Band (1.530-2.700GHz). MSS services are provided, for instance, by Inmarsat, for ship-to-shore and aircraft communications.

By far, most fixed commercial satellite services use C- and Ku-Band today, although efforts to exploit the Ka –Band with new satellites are growing in commercial use in the Americas and Europe.

Broadcast Services

Satellite technology has been a key enabler in the growth of the television broadcasting industry around the world. TV channels broadcast by satellite totaled nearly 14,000 channels in 2004. Over 7,000 of these were broadcast through DTH video package platforms, with 4,700 transmitted on various broadcast, direct, and cable satellite networks. Revenues generated by TV programming from satellite TV subscriptions exceeded $15 billion in 2004, and have continued to increase as households around the world go on taking subscriptions.[3]

Over-the-Air (VHF/UHF) Network Broadcast Distribution

Today most major national broadcast radio and TV networks take advantage of the economics of satellite technology when it comes to point-to-multipoint distribution. Satellite is the standard means to distribute national signals from a central television studio, broadcast, or master control center to geographically dispersed local TV stations. Typically, the studio or network origination center uplinks a digital satellite signal or multiple distribution signals to its affiliated local TV stations. The local stations downlink the programming and redistribute it over-the-air to consumers' VHF/UHF TV sets.

In some cases, local TV stations may also have uplink capabilities and share programming with other stations — for instance for news originating in the region — or with the central or "Network" studio. In the U.S., for example, ABC, NBC, and CBS headquarters and uplink centers are in the New York City and New Jersey area, while Fox Broadcasting uplinks from its Los Angeles center.

Syndicated Broadcast Program Distribution

In addition to the ABC, CBS, NBC, and Fox Network signals in the US, TV stations often purchase and broadcast programming acquired from program syndicators such as Warner Bros., Disney's Buena Vista Television, Viacom's Paramount Television, and Sony Pictures International Television.

Traditionally delivered via analog feeds in North America so that the widest possible range of stations can downlink and record or play-to-air, the technique for satellite-delivery of syndicated programming now includes electronic video file-delivery solutions using IP technology.

Storing a day's worth of 5-minute news clips requires some server space, but far less space than is needed to store a complete day's worth of syndicated shows. Nevertheless, with declining costs, long form content store and forward over satellite has increasingly replaced tape-based program capture and storage. An example is Pathfire's service for Warner Bros. Pathfire signed a deal in 2003 with Warner Bros. to deliver syndicated programming to the broadcasters' 835 TV-station clients via satellite. Pathfire servers were installed at the affiliate TV stations to replace traditional analog TV feed transmission and tape playback.The service was implemented to cut Warner Bros. distribution costs by using less bandwidth while making it simpler for stations to prep the content for air.

Broadcast Contribution: Flyaways, Transportables and SNGs for News, Sports and Events

News and Sports – Contribution & Backhauls

As noted previously, satellite services are also used for contribution applications, or delivery of remotely produced live or non-live materials back to the studio or network control room for contribution to the finished broadcast. This includes both Satellite News Gathering (SNG), and backhaul, or transport from outside of the studio locations such as sports venues back to the television studio.

Satellite Newsgathering (SNG)

Satellite News Gathering is the process for delivering 'live' and 'breaking' news from the field to the broadcast studio for live or delayed integration to a news program. The method is used for fast-breaking news stories, sporting events, political events, and live-on-air interviews.

Satellite uplinks are used for both television and radio news reporting.[5] The ITU defines SNG as: "temporary and occasional transmission with short notice of television or sound for broadcasting purposes, using highly portable or transportable uplink earth stations." Many broadcasters around the world own SNG units. Press and news agencies like CNN, Reuters, AP and others, both own and rent SNG services for news, sports and other live TV events.

"Flyaway" terminals can be packed and shipped in an airplane or helicopter for remote newsgathering and reporting.

Companies like PACSAT, Global Link, PSSI Global and many

Satellite Services

others, offer SNG services on a rental basis in the US. GlobeCast, with its fleet of 50 SNG terminals in Europe and Asia, and T-Mobile in Germany are examples of international service providers.

SNG Equipment

The interior of an SNG van includes audio/video satellite uplink and downlink equipment and coordination circuit monitoring and control gear. The equipment may provide for data transmission and should be capable of being set up and operated by an advance crew of no more than two people within a few hours or less.

Transportable earth stations not part of a vehicle can also be used for newsgathering and considered SNGs. SNG sound and low-bit rate video may also be operated on a laptop-size, mobile-satellite service like Inmarsat's BGAN service. Today, some newscasts have even used video clips shot using cell phones, as long as they are of sufficient quality.[5]

The earliest SNG equipment used analog modulation, similar to what was used in the conventional Ku-band for a fixed uplink location. The vehicle also contained equipment to support TV uplinking, such as video exciters, upconverters, HPAs, and receivers. During the 1990s, digital modulation was introduced bringing numerous advantages due to reduced signal bandwidth and uplink power requirements, and gave rise to the technology of Digital Satellite News Gathering (DSNG).

The ND SatCom SkyRAY Light Antenna subsystem can support SNG as well as disaster recovery and business continuity applications (Photo courtesy of NDSatcom).

State-of-the-art DSNG vans today can deploy practically anywhere in the world. Signals are beamed between a satellite and the van, and between the satellite and a control room run by a broadcast station or network. Increasingly, Internet Protocol (IP) is being implemented for data networking.[6]

The rack size and power usage of gear like amplifiers, modulators, and video encoders has shrunk over the years, while satellite bandwidth efficiency has increased. A digital 9 MHz slot of bandwidth, for instance, can now deliver what would have taken a full 36 MHz slot of bandwidth using an analog transponder in the past, and is routinely used in many regions. Equipment is also being increasingly deployed which employ 8PSK modulation, increasing the bandwidth efficiency for digital traffic by up to 33% compared to the QPSK modulation previously used.[8]

Using advanced video compression techniques such as MPEG-4, digital video encoders available in the market today can now deliver acceptable quality High Definition programs in as little as 7 Mbps and Standard Definition programs in as little as 2.5 Mbps, offering additional 40-60% bandwitdh improvements over what can be achieved with MPEG-2 compression.

An SNG system typically features a roof-mounted Ku-band uplink antenna and RF system. (photo courtesy of Global Link)

With the growing demand for High Definition transmission, HD-capable SNG equipment and trucks are now increasingly the norm in trucks used for sports and events coverage.

Sports and Satellites

Television has been integral in the growth of professional sports as an industry, and satellite services have played a critical role in the broadcast chain bringing live sports events from arenas and venues to TV screens.

Users of satellite services for sports include:

- Broadcast Rights holders and Non-Rights holders
- Leagues and Federations

- Sports Agencies
- Production Companies

Uses of satellite transmission for sports include these applications:

- Broadcast Contribution of shared or global feeds
- International distribution
- Pay Per View (PPV) and DTH distribution
- IP encoding and webcasting
- Venue backhaul
- First and last mile connectivity
- Secured delivery to betting establishments

Sports Coverage

As fiber points-of-presence have been increasingly installed in major sports arenas, SNG-based satellite services have yielded market share to lower cost terrestrial point-to-point links for football, hockey, baseball and basketball games. Nevertheless, in the U.S., many sports venues are not able to deliver sufficient Standard Definition and High Definition video signals via fiber, and satellite remains the best and only option.

Ad Distribution

In the U.S., companies like Pathfire and Williams Vyvx Services began beaming commercials over satellite to TV stations in digital file format in the 1990s. The typical 30-second spot, MPEG2-encoded at 8 Mbps, takes 30 MegaBytes of storage on a server. With today's reduced cost of server and storage, caching server costs are in the few thousand dollars–versus tens of thousands a few years ago–for an off-the-shelf service that can hold 1000 spots. Satellite multicasting is an ideal way to refresh hundreds of widely dispersed server sites rapidly with the same spot files.

In another example, Dallas-headquartered DG Systems used a Ku-Band satellite network to deliver advertising spots to nearly 3,700 TV broadcast facilities around the country for its client base of 5,000 advertisers and agencies.[7] In 2005, according to DG Systems, they were the

The Satellite Technology Guide for the 21st Century

Case Study: Covering the Olympics

With its cumulative worldwide audience of over 2.5 billion people, US $1.498 billion in TV rights costs, and its universal appeal, the 2004 Summer Olympics in Athens, Greece is a perfect case study in the role satellites play in delivering international sports events. The European Broadcasting Union (EBU), an association of 71 public and national broadcasters in 52 countries in Europe, North Africa and the Middle East, doubled its full-time capacity on the Eutelsat satellite system during the Olympics. The EBU booked the equivalent of 306 MHz of Ku-Band capacity on Eutelsat's ATLANTIC BIRD™ 3 and e-BIRD™ and pre-launch capacity on the then just-commissioned W3A spacecraft, at 7°East. In addition to contribution circuits to affiliates, the EBU delivered hundreds of hours of live event coverage. The EBU set up an infrastructure dedicated to the event, deploying a new uplink site to deliver a total of 36 nonstop program feeds. NBC had flyaways. CBS found a home away from home for its two dedicated live positions with views of the Olympic Stadium, dedicated flyaway uplink facilities, and office modules at the a special facility provided by the SNG and event transmissions provider GlobeCast, the world's largest provider of satellite transmission and production services for sport broadcasting. CNN also had separate live positions, a dedicated flyaway uplink and workspace provided by GlobeCast as well.

Broadcast Center for the 2004 Athens Olympics.

largest single digital distribution network for the delivery of television spots, with a network reaching nearly 95% of the top rated TV destinations in the top 100 U.S. markets.[8] DG Systems installed its Digital spot box receiver at TV stations, enabling the network to digitally receive ads from agencies for insertion of national spots in its programming.[9]

Internet Protocol (IP) and Broadcasting

Distributed server networks connected via satellite are playing a growing role in broadcast news operations.

Roswell, Georgia-based Pathfire, Inc. developed a satellite-based IP (Internet Protocol) multicast store-and-forward content delivery system to digitally transmit broadcast-quality video to stations. The system has been used by NBC to deliver its news channel affiliate feed service, and it was deployed for ABC's News One. The system allowed local station news producers to receive, sort, and order clips of edit-ready format footage from their desktops, avoiding the hassle of satellite downlink feed tape recording and tape-based operations.

In the same manner, CBS Newspath, which provides news services to over 200 CBS stations and affiliates, rolled out a digital file-based solution for its owned and affiliated stations. CBS used digital IP store-and-forward technology to reduce costs by minimizing satellite-refeeds, eliminating tape equipment and automating news production processes.[10]

Cable TV Distribution

In 1975, Home Box Office (HBO) inaugurated a new era. It leased a C-Band transponder for $1 million per year on RCA's Satcom I, the U.S.A.'s second commercial satellite. This allowed HBO to deliver television programs via satellite to cable systems which paid $10,000 to install 3-meter (9 feet) wide dishes to receive C-Band signals.[11]

Following HBO in 1976, Ted Turner acquired a transponder and uplinked his UHF station in Atlanta, creating WTBS. He nicknamed it "America's Superstation," offering it nationwide to cable systems.

It soon became the industry standard in North America, Europe, Asia, and eventually the rest of the world to feed TV network signals to cable systems via satellite. Extensive terrestrial microwave networks were rapidly replaced by satellite technology while satellite operators experienced rapid growth in demand as multichannel TV expanded the market.[14]

In 2005, nearly 9,000 cable systems would provide subscription programming to 67% of U.S. homes[15] with 339 national channels,[16] mainly delivered via satellite.

The Satellite Technology Guide for the 21st Century

Radio Distribution

Satellite is also used to broadcast network and syndicated signals to radio affiliates. Radio distribution via satellite is similar to TV distribution to cable television head-ends in that a large number of radio affiliates (1,000s of sites) may receive syndicated and network program feeds via a satellite transponder. For example, more than 8,000 radio affiliates across the U.S. depend on SES Americom satellites for programming feeds and uplink/downlink services. Analog and digital services for both C-Band and Ku-Band applications are provided to radio programmers. National, regional and specialized networks also benefit from Single Channel Per Carrier (SCPC) services.

In 2005, SatcoDX counted over 4500 digital radio channels on satellite around the world, with only 118 remaining in an analog satellite format, the majority of remaining analog signals being over Europe.

As terrestrial multicasting bandwidth costs and reliability improve, satellite-based distribution of radio to AM/FM station transmitters will continue to shrink. However, Direct-to-Home satellite radio services like XM Satellite and Sirius in the U.S. compete directly with AM/FM radio for listeners and will generate far more revenues and channels for the satellite industry than it derived by providing the backbone for national radio signal distribution.

Occasional Use Services

Occasional Use services are used for "one-off," ad hoc and special events, breaking or daily news, media tours and other program transmissions. Occasional services are typically provided on capacity that is available for a 30 day "booking window." The owner of the transponder usually has a process where the occasional capacity can be sold off to a full time user on 30 days notice. As a result, ad-hoc or occasional space is usually sold subject to scheduling availability, first on an inquiry, and then a firm-booked basis, with penalties if the time is not used.

Teleports

Uplink services can also be leased or purchased from satellite transmission service providers who operate teleports. Teleports provide terrestrial network interconnections to satellites, usually offering full-time and occasional uplink and downlink services accessing multiple satellites.

Satellite Services

By relying on a teleport to uplink, a TV network can concentrate on its programming, ad traffic, and sales, and production operations, and avoid business challenges such as:

- Capital investment in satellite uplink, video encryption and compression systems;
- Investment in human and operating resources and real estate;
- Requirements for satellite zoning and transmit licenses; and
- Time to build a satellite broadcast uplink facility.

Rather than expending up to US$ 2 million in capital and establishing 24/7 network operations, monitoring and other functions, many corporate and video users of satellite service rely on the services of teleport service companies.

Teleports traditionally derive most of their revenue from uplinking and space segment resale. Uplink services are typically offered on a full-time per-channel basis for video signal or data/Internet traffic, as well as on an "ad-hoc" or "occasional-use" basis for broadcast television requirements.

Cable, DTH (Direct to Home), Broadcast Distribution and Contribution signals are often uplinked at teleports on behalf of client programmers.

Most video uplinking by teleports involves transmission of compressed MPEG format video and associated audio streams in either an SCPC (single channel per carrier) or MCPC (multiple channel per carrier) mode.

International teleports provide Standards Conversion of incoming and outgoing international video signals between NTSC in North America and Japan, for instance, and PAL, or SECAM standards used in other countries.

In summary, teleport functions and services include:

- Uplinking and Downlink signals
- Data and video encoding & multiplexing
- Satellite and terrestrial circuit interconnection

- Data protocol conversion, and video conversion
- Signal monitoring & troubleshooting
- Video & Audio parameters monitoring
- Data and MPEG video transport stream analysis and reporting
- Hosting, Maintenance and Monitoring of customer-premises video equipment
- Fiber and terrestrial circuit monitoring

Downlinking

Downlinking is the reception of satellite signals using a receive system. Typical receive systems consist of the outdoor equipment – the antenna and feedhorn, Low Noise Amplifier or Low Noise Block downconverter – and IF (Intermediate Frequency) cable to the indoor equipment. The indoor equipment may be a series of devices including a Downconverter and Decoder such as an IRD (Integrated Receiver Decoder), or an integrated system connecting the devices.

Downlink services can be purchased on a full-time or part-time (hourly) basis typically from teleports and service providers. In such a case, a particular satellite signal downlink may be ordered and delivered to a particular customer circuit, or to a public switching point or "meet-me" point.

Conditional Access Management

Pay and subscription cable and satellite channels have stringent security requirements to protect their signals from unauthorized access by unwanted satellite receivers. The video uplink can house an encryption system, typically a computer network, or receive encryption information sent from a remote location such as the broadcast Network Origination point.

Teleports can provide Conditional Access Management as a service to client networks, authorizing specific clients sites access to the signal. This typically consists of inputting the correct data codes into the Conditional Access Control system, and ensuring that the authorization data is correctly inserted into the final outbound signal. A database of authorized receivers or users is maintained and the service provider coordinates by phone, email, or fax with the remote sites to activate, or deactivate the IRDs (Integrated Receiver Decoders) as required. Strict security controls, sometimes including technology export controls must be applied to protect systems. Coordinating the distribution of Security code updates, patches, smart card, and CAM (Conditional Access Model) replacements are other functions that may be required to manage an ongoing network and prevent security comprises.

IRD (Integrated Receiver Decoder) Authorization Services

Teleports may manage encryption authorizations for a channel full-time, or for specific broadcasting events. For instance, many corporate 'Business Television" events may include sensitive corporate internal information, and thus are encrypted to ensure a company's authorized sites receive the information, but competitors do not.

Turnaround Services

Downlinking a signal from one satellite, and retransmitting it to a transponder on another satellite is known as "Turnaround Service." Turnaround services are required to relay international program events around the world. For instance, a "live shot" feed of the summer Olympics in Athens may be turned around at a US East Cost teleport to a Domsat, or domestic US satellite. The Domsat then retransmits the feed for direct reception at broadcast affiliates across the U.S., which are not equipped to receive the international satellite feed. In this example, the satellite connecting Athens to North America would only be visible from the East Coast and require larger earth station antennas not available at a normal TV station.

With an IF (Intermediate Frequency) turnaround, the downlink signal IF output of the downconverter is directly input to the uplink satellite upconverter, avoiding the need to demodulate and decode a signal to baseband and then re-encode and modulate the signal again. This may save equipment requirements at the teleport for encoding and modulating equipment. Nevertheless, downlink demodulating and decoding still may be required for Quality Control and "Confidence" monitoring, where personnel at the teleport watch the signal on a TV screen and ensure the quality of the signals being transmitted to the satellite and relayed back to earth.

The Satellite Technology Guide for the 21st Century

Digital turnaround refers typically to passing a downlink signal from one satellite to an uplink on another satellite, while keeping the signal in the digital domain. For example, an MPEG-2 standard Single Program Transport Stream may be output from a downlink demodulator's ASI interface and input to the ASI interface of a digital modulator feeding the uplink to the other satellite. In this example, the MPEG-2 stream is not decoded to a baseband analog or digital SDI video, and then re-encoded for transmission to the second satellite. As a result, picture quality reductions that can occur as a result of multiple generations of decoding and re-encoding can be avoided.

Playout & Recording

A common service provided at teleports is Tape Playout and Records for broadcast programs. A professional Video Tape Recorder (VTR) is employed to record a downlink off a satellite so that the client can pick up a tape of what was broadcast at a later time. With a playout, the client tape is delivered to the Teleport, which is responsible for playing the tape in a VTR and sending the signal(s) output from the VTR to an appropriate communications path or satellite.

For example, a reporter covering the Cannes Film Festival in France for a show like Entertainment Tonight may be required to deliver footage of his interview with a movie star in time for the evening's broadcast news show. The reporter orders a playout and satellite feed from a service provider at the Cannes Film Festival who plays the tape into a satellite feed from France to US. The footage is then routed to the production studio in Hollywood, and recorded to tape for later editing into a show.

Time Delay Services

While tape records/playouts are often still used for time-shifting broadcasts, increasingly video server technology is becoming more efficient and cost effective for time-shifting. An example is GlobeCast, which for its WorldTV DTH service brings in channels from 23 world time zones to its Los Angeles Media Center. To provide channels from Africa, for instance, with a time-delay service, Globecast's practice has been to record the Africa broadcast day arriving via an international satellite and fiber link. It then adds a delay of several hours to the program signal so that the morning news from Africa is broadcast-in the morning in America. A

> ## Selecting a Teleport Vendor
>
> Considerations in selecting an uplinker may include:
>
> - Connectivity to terrestrial fiber networks locally and around the world. Availability of uplink antennas to the desired satellite(s).
>
> - Availability of Ancillary Non-Satellite Services such as Program Origination, Post-Production, and TV Production, IP and data services.
>
> - Proximity of the Teleport to the signal origination location, and cost of bandwidth to connect to the teleport from the customer site.
>
> - Redundancy & Backup systems: there should be sufficient UPS power, and access to backup power generators in case of utility power outages.
>
> - Diversity and Backup Facilities: Larger providers can re-route traffic from one teleport to another via fiber circuits, for instance during local disasters or emergencies. Diverse physical routing of fiber circuits into the teleport building diversifies against the risk of outages due to a terrestrial line cut or outage.

rule of thumb about time delay services is "the more you delay, the more you pay" since increased server storage capacity is required.

Network Management Services

Another service function that satellite transmission companies provide is Network Management. Because most full-service teleports are staffed with trained technicians on a 24/7 basis and equipped with signal diagnostic equipment and network status reporting equipment, satellite service clients may rely on a teleport to watch and supervise the program or data network signals. Teleports can also coordinate the activation and deactivation of equipment tied to a satellite network operating through uplinks at the teleport.

Help Desk Services

Help Desk services are an ancillary function that may be desired in conjunction with the operation of a satellite multi-site network for corporate enterprise applications. It can be outsourced, particularly in corporate networks. Typical "Help Desk" functions include tech support help line assisting client sites with receiving a broadcast or signal. As an example of this service, a company like Hewlett Packard might use Help Desk services from GlobeCast Enterprise Services to support the logistics and hookup of hundreds of offices around the world as they tune into HP's internal global company satellite broadcasts.

Trends in Teleports

As traditional broadcast applications implement technologies using the Internet Protocol (IP) in their production, contribution and distribution activities, Teleports are increasingly being required to provide IP-based services around the world.

As a result, the teleport needs to manage IP routers and access devices and add connectivity to Internet backbone providers as part of its menu of capabilities, even to serve broadcast clients. For example, Television ads are now increasingly delivered in parts of the world as digital files via the Internet to a TV channel's Master Control and Network origination center, which may be co-located at an uplink center. The availability of IP connectivity and secure environmentally controlled technical space for hosting of customer equipment such as IP servers, firewalls, encryption and other gear is another consideration.

Telecommunication Services

Data Services

Data services comprise the second largest application after video for satellite. Historically, satellites were used to provide dedicated international point-to-point leased line bandwidth for voice and data. However, as fiber networks have been implemented around the world, satellite services have been relegated to thin-route locations without fiber, or point-to-multipoint applications where the cost of satellite bandwidth and networks is competitive compared to the price of terrestrial solutions. Data broadcasting services and interactive services using VSAT technology are covered in detail in the next chapter on VSATs.

Satellite Services

Data broadcasting services that use satellite include:
- Financial & business data broadcasting – stock market news and information
- Connecting cell tower networks
- Two-way broadband access and Internet "cyber cafes"
- Remote office and SOHO (Small Home/Office) connectivity
- Remote location Internet Access
- Rural telephony via SCPC or VSAT networks
- Remote wireless/cell connectivity for rural
- In-Store Music broadcasting – such as provided by Muzak
- Digital Advertisement Delivery
- Delivery of Paging data to Paging transmitters
- ISP Backbone Connectivity for remote ISP Points-of-Presence

Distance Learning

Satellites are also used to broadcast lectures via close-circuit video networks to dispersed learning sites. Combined with voice or interactive data keypads, video networks help students in different sites learn and interact, though they may be in classrooms thousands of miles apart, or at home. Rural communities and developing countries have used satellite distance education to implement national learning initiatives and nation-building activities.

Corporations can also benefit from distance learning by satellite by avoiding the costs of travel involved in bringing employees from far-flung locations to a single meeting place. The cost savings from avoiding travel, time away from performing a job, as well as personal time away from family, are key factors that make satellite-based distance learning a good investment for many companies.

Telemedicine

Another benefit provided by satellite technology is in "telemedicine," where experts on a specific topic can provide a lecture to a group at a convenient time. For example, some surgeries have been performed by a lecturing surgeon while medical students located thousands of miles away obtain valuable medical training. Students and doctors may even participate in remote medical consultations using interactive satellite video/data channels.

Direct-to-Home (DTH) Services

Satellite TV

Stanford Professor H. Taylor Howard pioneered the technology in 1976 to receive television signals in the home directly from satellite broadcasts. Today, the user at a home purchases a pizza-size satellite dish and directs the dish to a satellite in the sky. DTH technology has evolved together with local and cable television. Scrambling and descrambling technology was added to extract payment for premium content, and digital broadcasts replaced analog broadcasts to improve visual and audio quality. The size of satellite receiver dishes has decreased from over 12 feet to about a foot and half (45 cm) in diameter over the years. DTH has become a serious rival to cable television, as there were nearly 30 million DTH subscribers in the US in early 2007, or approximately 1 out of 4 subscribers to multichannel video program distributors. Leading the way were DirecTV with 16 million subscribers and Echostar with 13.4 million subscribers as of March 2007.

In England, the BskyB platform, beams almost 400 channels on four Astra and Eurobird satellites to 17 million viewers in the United Kingdom.

In the United States, DirecTV and Echostar's DISH network each operate multiple satellite orbital positions, with hundreds of channels available to consumers using 18" antennas and set-top receivers. These can be also purchased with TiVO™ or Digital Video Recording (DVR) functionality, giving satellite the "on-demand" capability. The third largest US operator, GlobeCast provides the WorldTV DTH platform in the Ku-Band, which by early 2007 delivered over 200 TV and radio channels from around the world in native languages from Europe, Asia, the Americas and Africa.

Astra and HotBird satellites in Europe are the major systems for regional DTH distribution, carrying a combined 2500 channels. The Astra system reached an estimated 102 million homes in 35 countries, while HotBird satellites reached 110 million subscribers in over 150 countries.

DirecTV

DirecTV was founded in 1993 by Hughes Electronics, a subsidiary of General Motors. The company increased its satellite capabilities by acquiring USSB in 1998 and Primestar in 1999. Moreover, DirecTV grew its subscribers from 320,000 in 1994 to 16 million in the first quarter of 2007. In 2003, the controlling interest of DirecTV was sold to News Corporation, which then transferred its shares in DirecTV to Liberty Media in late 2006. DirecTV offers DTH services via its ten satellites over the United States, the first launched in 1993.

In mid-2007, the regular monthly fees for basic, plus, and premium service were US $49.99, $54.99, and $ 59.99 per month, offering over 140, 185, and 185 channels, plus the use of Digital Video Recorder (DVR) technology, respectively. In addition to the monthly fee, the user must employ both a satellite dish and a descrambler box, which is sometimes included in the service package. Common dishes include the 18" dish and the 18"x20" dish to receive multiple satellite signals. Finally, premium services such as DVR and High Definition TV further increase the monthly fees and require additional equipment.

Echostar's DISH Network

Echostar was founded in 1980 as a distributor for C-Band TV. Operating with its eight satellites over North America, first launched in 1995, Echostar has increased its subscribers from 100,000 in 1996 to 13.4 million in early 2007. During the company's history, Echostar was the first service to provide Digital Video Recording (DVR) and the first to offer local programming in all 50 states.

In order to view programming, the user must purchase a receiver, ranging from the most basic at US $ 90 to the most premium (able to receive HDTV broadcast and with recording capabilities) at US $ 700. Furthermore, Echostar provides a variety of content packages, from a basic 60-channel plan at $29.99 per month to an all channels plan (over 230, including the premium movie channels) at US $89.99/month.

The Satellite Technology Guide for the 21st Century

Satellite Radio

The first made-for commercial radio satellite system was initiated by WorldSpace, Inc., which started with the goal of providing radio to the masses in Asia and Africa using low-cost shortwave radio-like custom-built receivers capable of receiving signals from WorldSpace satellites.

Satellite radio services in the U.S. rapidly gained subscribers, with some 7.7 Million XM Satellite Radio customers, and 6 million Sirius satellite subscribers at the end of 2006. Both companies launched spacecraft and services in 2002, requiring subscribers to purchase a satellite radio (initially for use in cars) and a monthly subscription to mostly commercial-free radio channels programmed to a variety of radio listening tastes. Both companies, like their predecessors, DIRECTV and Echostar, have spent hundreds of millions of dollars marketing their brands and offerings to consumers, including large sums on exclusive programming rights.

XM Satellite reportedly signed $650 million for an 11-year contract right to Major League Baseball on satellite radio, [13] while Sirius made waves when it inked a $500 million, five-year contract with syndicated radio "shock-jock" Howard Stern, $30 million for Martha Stewart programming, $107.5 million for broadcast rights to NASCAR racing, and $220 million for 7 years of NFL programming.[14]

Direct-to-Home Broadband

Wild Blue

Wild Blue offers an Internet connection via satellite to the 15 million homes and offices that are not wired for cable or DSL Internet access in the U.S. Launched in 2005 and the first true commercial Ka-Band satellite service venture in America, Wild Blue offers a Basic package of downloads at 512kbps and uploads at 128kbps via a 26-inch dish and modem. Prices in 2006 started at US $49.95 per month plus US $299 for a 26" satellite terminal and $179 for installation. Higher end packages such as Select ($69.95) and Pro ($79.95) offered download speed increased to 1Mbps and 1.5Mbps, with upload speed increased to 200kbps and 256kbps, respectively. Like most ISPs, Wild Blue also provides email and web hosting services.

Satellite Services

The Wild Blue infrastructure includes the Telesat Anik F2 satellite over the US (launched 2004) and five satellite gateways that connect the Internet to the satellite. Service initiated in June 2005, and the company plans to launch a second satellite based on the demand of the market. The company attracted various satellite investor groups including Intelsat, the National Rural Telecommunications Cooperative (NRTC), Kleiner Perkins Caufield & Byers, Liberty Media, and Telesat.

DirecWay

HNS' DirecWay service is another two-way broadband satellite Internet solution available in the contiguous United States. DirecWay served 250,000 consumers and small businesses by mid 2005. DirecWay targets customers from the 15 million American households and businesses that are underserved by cable and DSL, but unlike WildBlue, uses conventional Ku-Band transponder capacity.

New and Emerging Services

Satellite technology is the most cost-effective solution for one-to-many broadcasting - be it TV or data. New services and emerging applications include Telco TV, PVR services, and Interactive TV, High Definition TV, and Internet TV, satellite delivered mobile TV, and more.

As broadcast and cable network, program, spot, and promotional programming contribution and distribution are increasingly being managed in digital file formats while storage costs continue to decline, television operations are increasingly using satellite enterprise networking technologies.

Video-On-Demand (VOD) Distribution

A key competitive feature of cable TV versus DBS is cable's ability to provide true video-on-demand (VOD), which DBS cannot offer. Cable operators have been steadily rolling out commercial VOD services in key systems, after over a decade of technology development and trials. As VOD has rolled out to an increasing number of head-ends sites, satellite multicast technology has become increasingly cost-effective for VOD file distribution, replacing tape-based delivery.

Burbank-based TVN Entertainment, Inc., for example, manages delivery of over 3,000 programmer-hours per month of VOD content to cable systems, delivering encrypted movie files via its C-Band digital satellite network on Intelsat's Galaxy satellite system. TVN's offering has been deployed in top cable companies, with nearly 100 sites around the US receiving satellite VOD content. In 1996, about 1500 hours per month of refreshed data was sent, and some affiliates took down as much 600 hours of content. At that time, all of the VOD content could be multicast over a single satellite transponder via a 13 Mbps of data capacity and delivered to docking stations located at the cable head-end. Likewise, Pay-Per-View programmer iN Demand is providing VOD offerings featuring studio and sports content delivered via its satellite network for Affiliate VOD services.

Using TVN's asset management software, movie content suppliers can remotely track their on-demand content assets through the delivery system. The docking station rebuilds files sent over the satellite until they are complete as a pre-described package consisting typically of a movie, or trailer and metadata file. Once the package is complete, it is pushed over to the VOD server system's asset manager. A web utility allows affiliates and content providers to view the database to follow the status of the asset.

HDTV

High Definition Television (HDTV) brings stunning, cinema-like picture quality and high-fidelity surround sound audio to the TV viewing experience.

According to a Kagan Research study released in 2005, HDTV subscribers were forecast to grow at nearly 30% per year over the next decade to nearly 94 million by 2015 in the U.S.

Because High-Def signals consume over twice the transponder capacity as Standard Definition TV, the growth of HD channels for cable and satellite in recent years has boosted demand for cable and direct-to-home capacity on satellites. For example, SES Americom and Intelsat both established competitive "neighborhoods" on their cable TV satellites for HD distribution. DirecTV has acquired two complete Ka-Band satellites originally intended for broadband Internet access in order to deliver both local stations over DBS and HD signals in the future.[15]

Interactive Television

In order to compete with cable TV, DTH providers have introduced subscription interactive services, such as SkyActive on BskyB's

Satellite Services

service in the UK. SkyActive provides broadband information and entertainment, games, and contests, sports highlights and more, accessible on demand by consumers via their remote. Interactive services provided via DBS are essentially data-cast services via the satellite. While they offer additional subscription revenues, the amount of data broadcast to consumer is negligible compared to the amount of data required to deliver video channels.

IPTV Ventures

IPTV, or the distribution of Internet Protocol-based television services over broadband networks, is the newest competitor to traditional cable TV. Communications research firm Multimedia Research Group, Inc. (MRG, Inc.) projected in 2005 that the global IPTV market would grow from 2 million subscribers in 2004 to 26 million in 2008, but the market has not blossomed as quickly as analysts had originally forecast.

In Europe, Asia and now in America, phone companies providing high-speed broadband, ADSL and Fiber-to-the-Home services have entered the subscription TV market served by cable TV and DBS, by introducing competition in the form of subscription and pay TV over their Internet-capable infrastructures (or IPTV).

SES Americom launched IP-Prime in 2005, a centralized, satellite delivered IPTV distribution solution for telcos to bundle traditional standard and HD programming with voice and broadband services. SES announced at that time an agreement with the National Rural Telecommunications Cooperative, which serves rural and independent telephone operators in a market of more than 10 million homes to trial the service in 2006. IP-Prime originates from the SES Americom IPTV Broadcast Center in Vernon Valley, N.J., where video and audio are received and processed for distribution via satellite and fiber to telco video hubs nationwide.

In-flight Internet

Boeing Connexion was launched to provide high-speed Internet access (5Mbps download, 1Mbps upload) during airplane flights. The Connexion division was created in 2000, and testing of the in-flight service began in 2003. Boeing equipped planes with satellite transmitting and receiving capabilities, so that users could access the Internet, company Intranets, and their private email. Various satellites and ground gateways carry the transmitted information from and to the Internet.

The Satellite Technology Guide for the 21st Century

Although Boeing installed the equipment on its planes, it contracted various companies for the modem (ViaSat), satellite, and gateway capabilities in the Connexion system.

After entering into commercial use in 2004, the Connexion service was terminated at the end of 2006 due to disappointing usage and revenues figures.

The failure of the Connexion system was not due to technical reasons but for failure to assess correctly the potential demand for such services. The Connexion experience is instructive of necessity for careful business planning when launching new services. Other companies such as ARINC and OnAir continue to provide in-flight Internet services albeit on a much smaller scale that Connexion. In-flight entertainment services have been succeful however, as air carriers such as JetBlue have helped to brand their "experience" by including DirecTV on board the aircraft for travellers.

Digital Media Broadcast (DMB)

The S-DMB (Satellite – Digital Media Broadcast) system employs satellites to broadcast multimedia directly to handheld mobile terminals. In Korea, where the buzzword "cellevision" describes the new service of providing television content to cellular phones, the service has caught on quickly. Manufacturers such as Samsung and SK Telecom began producing phones with multimedia capabilities, selling 1,500 DMB enabled phones a day on average in mid 2005. A joint venture between TU media and SK Telecom attracted more than 100,000 subscribers in three months.

TU Media's S-DMB service retailed in Korea for the equivalent of about US $13 per month plus a $20 activation fee in 2006. Subscribers were able to access nine video channels featuring a variety of sports, news, and music content, as well as 25 audio channels. TU Media claimed 1.1 million subscribers at the end of the 2006, and is projecting 5.5 million subscribers by 2010.

Conclusion

Satellite Services will continue to be the backbone of the satellite industry. New applications such as IPTV, Mobile TV and HDTV should be requiring new bandwidth and value added services that the satellite companies can provide. As media consumers' demand for these services and expectations continue to rise, there can only be more opportunities for satellite service providers in the new media environment.

Satellite Services

Notes

[1] Satellite Industry Association, Presentation, "State of the Industry Report", presented at ISCe Conference, June 2006.
[2] SatcoDX. http://www.satcodx.com/usa/
[3] Euroconsult report "TV Channels 2004."
[4] Jonathan Higgins, "The Future of Satellite News Gathering, www.braodcastengineering.com, May 1, 2005.
[5] Mark Brown, "Video Over IP for News Acquisition," www.broadcastengineering.com, Sept. 1, 2004.
[6] Philip Bird and Khalid Butt,, "Digital satellite news gathering: More features in less Space," www.broadcastengineering.com, July 1, 2002.
[7] http://www.dgsystems.com/about_us/corporate_profile.html
[8] http://www.broadcastbuyersguide.com/bbg/editorial/archives/dgsystem_20050206
[9] http://www.dgsystems.com/network/video_distribution.htm
[10] http://www.satnews.com/stories2/4july2001-7.html
[11] http://history.acusd.edu/gen/recording/television3.html
[12] http://www.ncta.com/industry_overview/indStat.cfm?indOverviewID=2
[13] Los Angeles Times, Sunday, July 10, 2005, "Rivals, and Friends", Page C1.
[14] ibid.
[15] "Global Assessment of Satellite Demand: A Demand-Driven, Region-Specific Analysis of the Commercial Geostationary Satellite Transponder Market for 2003-2009", Northern Sky Research (NSR).

CHAPTER 7

VSATs

by John M. Puetz

Since VSATs (Very Small Aperture Terminals) were first introduced over 20 years ago, over one and a half million interactive terminals have been put into operation worldwide and well over two million receive-only terminals. The technology has literally changed the way businesses and governments around the world conduct operations.

Virtually every industry segment that one can think of use VSATs to conduct all or a portion of their business activities: retailers, banks, brokerage firms, stock exchanges, cruise lines, energy services companies, construction firms, realtors, schools and universities, trucking companies, automotive dealers, utilities, travel agencies, railways, agriculture, and government agencies such as customs, postal services, police and military services. And over the past five years, over a half-million consumers have been using one and two-way VSATs to gain access to the Internet.

So why has this technology been so widely adopted? In short, VSATs offers benefits that no other communications medium can.

Foremost is satellite's ability to provide a uniform level of service over huge geographical areas. CIOs and IT managers for regional, national and multinational enterprises, struggle with the challenges of communicating with numerous business locations that are widely separated and have various degrees of supporting telecommunications infrastructure. The ubiquitous service coverage of VSATs coupled with uniform service capability and quality, are the primary drivers for dozens of retailers, automotive dealers and service station chains in the United States to deploy tens of thousands of VSATs per company. While the US is noted for its highly developed telecom infrastructure, broadband fiber, DSL or even leased line services are not widely available in every small town across the US. Until VSAT systems were deployed, the lowest common denominator service IT departments could count on was dial-up, which is woefully inadequate for any meaningful data service. For the past five years over 250,000 VSATs have been deployed in North America alone, which accounts for approximately 45 percent of deployments worldwide. VSATs' great success is clearly demonstrated in these high uptake rates that are occurring worldwide.

Another significant benefit of using VSATs is economic in nature. As you have already read from previous chapters, satellites have the innate ability to deliver information (data, stock quotes, video, movies, etc.) from one location to many locations (literally millions) with a single transmis-

Figure 7.1 - Cumulative International TDMA VSAT Shipments Worldwide

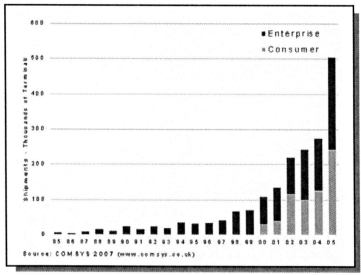

Figure 7-1 shows the annual shipments of commercial and consumer VSAT terminals. Over one-and-a-half million interactive (two-way) terminals have been shipped and placed into service over the past twenty years. And as can be deduced from the figure the vast percentage, over a million, have been placed in service in the past five years. And consumer based services, like WildBlue, DirecPC, and international based services, account for over half of these million.

sion. To date the Internet and most corporate intranets have yet to establish a multicast streaming infrastructure that permits this same one-to-many benefit. Currently each recipient of streaming content must make individual UDP connections to the streaming server. Satellites and VSATs have long been used for information distribution, especially critical real-time data like stock exchange quotations or television programming and live news feeds to cable headends and broadcasters.

Further to the economic benefit is that implementation and service costs/pricing are independent of distance—unlike any other telecom media. Additionally, the cost-effectiveness of broadband over satellite arises from the unique combination of broadcast capabilities within full networking solutions, without limitation of distance, geography or location. And VSATs have the ability to provide asymmetric services, thereby providing a best-fit match for content distribution, Internet access, distance learning and similar services.

Enterprises are finding that VSATs are not only economically competitive with other telecom alternatives, but from a total service delivery solution they can be unbeatable in many instances. Many VSAT service providers provide total turn-key managed network services to their clients, enabling enterprises to remain focused on their core business. A growing number of companies use VSATs to augment their existing wired facility network infrastructure, selecting only the "difficult-to-reach" or the "poorly-served" locations for VSAT services.

Another significant benefit is that VSATs are wireless requiring only electrical power to operate. They bring connectivity or broadband service to locations that would otherwise be without. Examples include Internet access, email and telephone services on cruise ships, deep-water oil drilling platforms, commercial vessels, remote mining sites, commercial shipping and even commercial and business aircraft. And since no existing infrastructure is required, a VSAT terminal can be rapidly deployed and made operational within minutes, as is often done in the energy services, construction, disaster recovery and news gathering industries. A network of a few sites or a few dozen sites can be established in days, not weeks, months or even years as is required for wired facilities. And scaling up or down the number of service locations or capacity is straight forward and easily accomplished.

Further, since VSATs provide wireless services across thousands of miles and do not rely on telecom infrastructure to provide service, they are ideal for providing a back-up business continuity service for existing wired-telecom facilities. When wired connectivity outages occur because of flooding, storms, construction breakages or other causes, data and voice services can be automatically and seamlessly routed over the satellite network.

Finally, an often overlooked benefit of VSAT networking is the network management infrastructure that is inherent with these systems. All broadband VSAT systems have a centralized control and management system, which makes network operations, maintenance and servicing a snap. Network performance, capacity management, service level agreement (SLA) and quality-of-service (QOS) auditing are no longer relegated to service wish lists.

20-plus Years of Heritage

Interactive VSATs found their birthplace in the United States (Southern California) during the early 1980s when the telecommunications regulatory environment eased and entrepreneurs jumped in with both feet. The earliest systems were one-way, taking advantage of the strongest benefit that satellite has to offer—distributing the same information to hundreds and thousands of locations. The initial data rates were very low by today's standards—2.4 Kbps which quickly grew to 19.2 Kbps—but fast for the time, especially when the only alternative was dial-up access at 1200 to 4800 baud. These systems were used to deliver financial information updates, news, stock quotations, meteorological data and weather forecasts, high resolution pictures, scientific data and remote electronic publishing.

To meet the marketplace economic demands, these receive-only terminals needed to be under US $2,500 installed. To achieve this, antennas were kept very small, less than 1.0 meter and typically 75 cm, and receiver electronics took advantage of new microprocessor and compact IC technology. But small antennas met that a specialized signal transmission approach was required to mitigate receive signal interference from adjacent satellites. Spread spectrum modulation was adapted from military use. As we'll read later on in this chapter, this technique literally spreads the information signal over a very wide frequency spectrum using specialized coding to avoid interference, noise and jamming. The added benefit of such a small antenna is that it was easily positioned, and in some cases, 75 cm antennas could even be located inside offices, provided they had an unobstructed view of the equatorial plane and the satellite of interest.

In the mid-1980s, two-way (interactive) terminals were developed. Since these systems also transmitted, larger antennas were necessary to minimize any transmit interference to adjacent satellites and to reduce transmitter power requirements for the VSAT. These interactive terminals used 1.2 and 1.8 M antennas and operated with a centralized hub facility which operated much larger 8 M to 12 M antenna. This large hub antenna allowed remote-to-hub return links to operate at low power levels, thereby also reducing the transmit power amplifier size and cost at the remote VSAT. Typically the outdoor electronics (antenna and RF) accounted for 50 to 60 percent of the cost of the entire terminal—this paradigm still holds true for today's VSATs.

While these original interactive systems seem archaic by today's standards—they were a marvel back then. They provided data services of 300 to 2400 bps per location, operating TDMA return links at 64 Kbps burst rates using a variety of data formats; SNA, HDLC, SDLC, and ultimately X.25. Data rates for the hub-to-remote (outbound) direction were originally 256 Kbps and doubled to 512 Kbps within a few years. For some end-users telephone service was provided for internal company business, connecting warehouses and key retail locations with corporate headquarters. Given the limited capacity of systems, voice services created a tremendous burden on throughput (consuming 25 percent of an inbound carrier capacity per call) even with the very latest in digital voice compression technology available at that time (ADPCM and RELP) operating at 16 Kbps. Needless to say, calls were kept to a minimum back then for this type of VSAT.

Most of these early systems were used for retail and automotive services, with early adopters like Sam Walton (founder of Wal-Mart Stores) and companies like Chevron and General Motors realizing that this new technology would play an important role in their business efficiency and success going forward. Most applications centered on inventory control, merchandise pricing and ordering, credit card authorization, and other back-office support applications.

Costs of the early interactive terminals were high, upwards of $15,000 for Ku-band and even higher for C-band when they were introduced in the late 80s. As can be seen from Figure 7.2, it took about six years for interactive VSAT prices to decrease by 50 percent to around $7,500 in the mid-90s. During this time period almost 100,000 terminals were shipped. While not yet significantly large, this increase in production volume enabled reductions in antenna prices, RF amplifiers and indoor equipment electronics, as did further technology innovation.

Figure 7.2 - Interactive TDMA VSAT Terminal Prices (1985-2006)

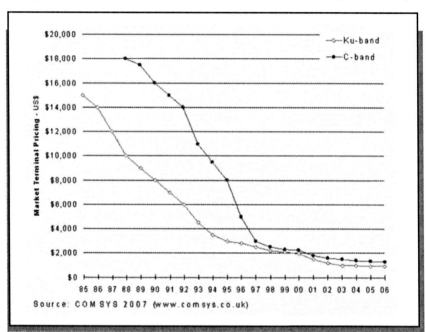

The early network management and control systems were built around mini-computer technology. In the case of Hughes Network Systems' (HNS) Personal Earth Station (PES), a VAX machine the size of several large business desks pushed together ran a custom UNIX based software package—no graphical user interface screens on those operator consoles—the concept hadn't been invented yet! Capacity management and network planning were cumbersome and time consuming using specialized computer based tools. These early systems required considerable resources to install, integrate into existing business computing networks and communications protocol infrastructures, and operate. Large companies that owned their own network and hub equipment hired and trained specialized personnel to operate the network once the system was installed and operational. Other companies turned to satellite service providers, who owned their own hub equipment, space segment and network operations center, and provided "shared hub" connectivity services for a flat monthly fee per remote location. For these network operators, putting together the earth station antenna, electronics and TDMA hub equipment was no small undertaking in resources or capital—costs of US $1.5

to $3 million were common place. Fortunately, prices have been dramatically reduced in the past several years. Today, several equipment manufacturers offer "mini-hub" packages that are under $200,000 and can support hundreds of remote sites.

Advances in technology, increased efficiencies in production, competition and significant increases in product volume brought further price reductions to interactive VSATs. By the late-90's the $2,000 barrier was being breeched for large network orders. By then, inbound (remote to hub) capacities had increased to 256 and 512 Kbps, while outbound capacity climbed to 1 Mbps. True broadband capacities were just emerging.

In the past five years, VSAT system capacities have become "broadband," with transmit rates per terminal of 256 Kbps to 1 Mbps, and receive rates at 10 to 30 Mbps. With the widespread use of IP (Internet Protocol) and the integration of IP routing and networking functions directly into VSAT terminals, service utility has increased substantially resulting in greater market acceptance. As a result system orders have exceeded over 100,000 terminals per year since 2001 for the non-consumer market.

Key to high data rates and lower terminal costs have been several technology innovations. The wide-spread adoption of the DVB/MPEG standard for television program distribution in the direct-to-home (DTH) markets around the world since the mid-90s has created millions of set-top boxes. These low-cost receivers (under $300) made very low cost receive (L-Band) front ends and DVB/MPEG demodulator/decoder chips and chipsets available for one and two-way VSAT use. Along with this technology came very low-cost (under $75) integrated RF receive electronics, known as LNBs (low-noise amplifier/block downconverter).

With RF components being such a substantial contribution to terminal cost, the introduction of lower-cost transmitters in the past five years, known as block up-converters/amplifiers (BUCs), operating with L-Band interfaces to the indoor electronics, have made a significant impact to the bottom line costs. Today, the cost of 1 and 2 watt Ku-Band BUCs are well under US $900 in reasonable quantities. This has made possible interactive VSAT terminals under US $1,000 in large quantities a reality.

Other general advances in technology, computers, electronic packaging and mass production have also had a positive impact on reducing equipment costs. Relatively inexpensive computer and digital signal processing (DSP) chips, field-programmable LSI chips, custom ASICs (application specific ICs), and mass memory are a few such examples.

Today, VSAT remotes are often referred to as "broadband satellite routers," have a footprint smaller than a laptop computer, and come with an integrated IP router that supports IP acceleration and the ability to receive streaming video. Voice, no problem—just plug in VOIP phones or connect to an IP-capable PBX. And IP video-conferencing works great too. VSATs from several vendors support virtual LANs (VLANs) so that network users from different organizations or businesses can be supported at one remote location (like an offshore oil platform, or a large multi-tenant office building). Each VLAN operates independent of the other thereby providing private network data integrity and security.

While we have focused on one-way and TDMA VSATs thus far, the oldest and simplest form of satellite communications is a dedicated circuit between two points. Otherwise known as SCPC (single channel per carrier), this technology has been the mainstay of satellite communications since the late 60s and early 70s when government sanctioned telecoms (PTTs) first used satellites to provide telephone and data communications. SCPC also applies to dedicated point-to-point circuits that utilize time division multiplexing (TDM) to aggravate multiple services (e.g., voice channels, data, etc.) onto a single satellite link. In the 70s, satellite modems (modulator-demodulator) entered the digital age and were reduced in size to occupy only 12 to 18 inches of rack space. Advances in error correction coding began with Andrew Viterbi's convolutional coding in the mid-late 70s. Higher levels of integration, coupled with advances in ICs, digital signal processing and error correction coding enabled modems to shrink in size (to 3 ½ inches of rack space) and price (US $5K to $7K) by the mid-80s while providing even better link performance and efficiencies.

By the late 80's complete SCPC VSATs (modem, RF electronics and antenna) were available for US $10K to $14K and had data capacities of 512Kbps to 2 Mbps. Further advances in modulation techniques and error coding enable modems to operate at data rates as high as 45 Mbps by the mid-90s. Integration of digital audio compression at rates of 128 Kbps to 256 Kbps enabled new applications in radio broadcasting and digital audio program distribution—even to commercial aircraft in the US by the mid-90s (USA Today/Sky Radio). With the advent of L-Band RF technology (BUCs) and L-Band modems in the late 90's SCPC VSAT pricing further decreased to the US $5K to $7K range. IP-based functions were integrated into SCPC modems in the early 2000's creating a new category, "IP modems" that provided IP acceleration, QoS (quality of service), traffic flow prioritization, and IP compression techniques. Continued advances in coding with turbo-product-codes, trellis and the newest irregular low-density parity check (LDPC) codes, coupled with 8-PSK and 16-QAM modulation techniques enable a one rack unit (1 ¾ inch) modem to operate at blazing speeds up to 238 Mbps.

MasterWorks estimates that between 15,000 to 20,000 SCPC modems and earth stations are shipped annually. This figure has been relatively stable over the past number of years and as result we estimate that there are 100K to 150K SCPC modems installed and operating worldwide (of both the open and closed network modem types).

Over the past twenty years, there has been considerable consolidation of equipment vendors, satellite operators and service providers. Of these three the equipment vendors have seen the most change. In 1997, likely the peak of the number of VSAT equipment vendors, there were 7 to 8 major vendors of interactive VSATs and some 15 major vendors for bandwidth-on-demand (DAMA) systems. Just six years later, the DAMA system vendors dropped down to five key players and currently there are now just three. Meanwhile the interactive VSAT (TDMA) system manufacturers actually increased slightly with the introduction of the DVB-RCS based systems.

Technology Overview

The term VSAT technically means very small aperture terminal, which refers to the antenna size. Generally speaking, antennas below 2.4m in diameter qualify as a VSAT, with Ku-band antenna sizes typically being 1.2m for commercial use and about 1m for consumer use (e.g., interactive broadband terminal). Figure 7.3 illustrates the major components of a typical two-way broadband VSAT terminal. The RF electronics is mounted on the antenna with an offset from the antenna reflector by 22.5 degrees to ensure that the equipment doesn't impact antenna transmit and receiver performance. Today, the cabling interconnecting the outdoor electronics (ODU) to the indoor electronics (IDU) typically operates at L-band (950 to 1450 or 1750 MHz) frequencies. This inter-facility link or IFL has separate cables for transmit and receive. The indoor electronics typically provides DC power for the receive LNB and transmit RF BUC and contains the satellite modem (modulator, demodulator, forward error correction coding). A number of manufacturers have opted to package all of the electronics (both indoor and outdoor) at the antenna location. While this approach consolidates packaging and elimi-

The Satellite Technology Guide for the 21st Century

nates the need for an RF-based IFL, the concept has had lukewarm acceptance. Practically all VSATs today incorporate IP functionality which may include, routing, protocol acceleration, header and payload compression, quality-of-service prioritization mechanisms and even caching capabilities.

Two broad VSAT category designations can be used to differentiate the target market segments for broadband services: consumer and non-consumer (or enterprise). The consumer marketplace has developed over the past 5 years with two-way Internet access services provided directly to consumer end-users by services like DirecPC and StarBand. The newer service offerings by WildBlue (in the U.S.) and more recently iPSTAR (in Southeast Asia) are considered second- or third-generation offerings. These VSATs are designed for very low-cost, optimized for Internet access and typically aren't suited for the heavier capacity and networking requirements of enterprise users. Compact mobile broadband offerings like Inmarsat's BGAN (broadband global access network) would fit more under the enterprise category, but these generally aren't considered VSAT terminals in the traditional sense—having both the antenna and baseband electronics in a single integrated package akin to a laptop computer, with the laptop screen back acting as a flat antenna.

Figure 7.3: Broadband-IP VSAT Terminal Major Components

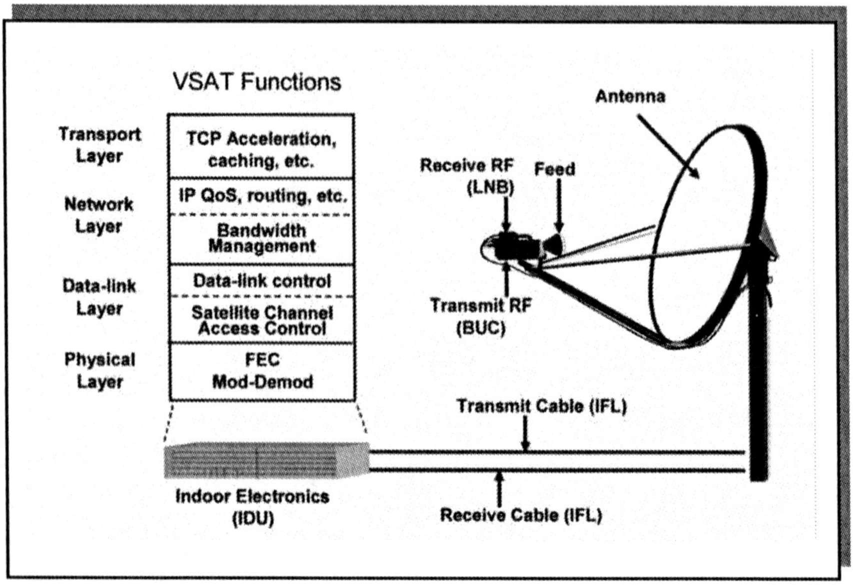

Source: MasterWorks Communications, 2007

Table 7.1 - VSAT Technology/Application Matrix

Connectivity/ Topology	Service Type	Capacity Flexibility	Satellite Technology	Best Fit Traffic Applications
Point-to-Point or Star	Circuit	Fixed (days/wks/mos)	SCPC	-High throughput circuit services -Dedicated Services -Static capacity
Star	Packet	every 1-4 secs	MF TDMA	-Bursty IP traffic -Moderate real-time traffic (VoIP) -Low cost hardware -Content Distribution -Keystroke interactive, point-of-sale, SCADA
Mesh	Circuit	1-5 min	FDMA (SCPC) DAMA	-Circuit services (e.g., voice, video conferencing) -Mesh Services -Well-defined traffic patterns -Low-moderate rate real-time traffic -Low multi-destination concurrency
Mesh	Packet	every 1-4 secs	MF TDMA	-Bursty IP Traffic -Mesh services -Any traffic patterns -Any rate real-time traffic (voice, videoconferencing, etc.) -Any multi-destination concurrency

Source: MasterWorks Communications, 2007

Technology Matches Applications and Use

Within the traditional two-way enterprise VSAT terminal category, there are four basic categories that represent all deployed systems. These categories are distinguished by the connection topology (e.g., point-to-point, star (or hub-spoke) and mesh), type of connection (circuit based or packet based), capacity flexibility (fixed or adaptable) and the type of technology used to access the satellite (SCPC, TDMA, FDMA, etc.). Specific details regarding connectivity and access technology are provided in the following sections.

As can be seen from Table 7.1 above, certain VSAT technologies are best suited for specific applications needs and traffic requirements. Because there is such a large range of requirements, including economic, there is no such reality as "one-size-fits-all". Providing services to a few locations that require high-throughput capacities that don't change much will be best served by SCPC technology. Why? Mainly economic—total equipmen costs and recurring operational costs are attractive, with very good space segment efficiencies.especially when the latest coding, modulation and waveform enhancement technologies are deployed.

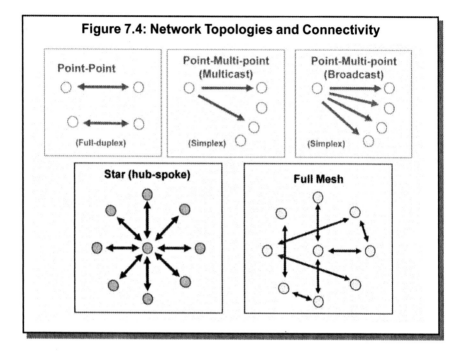

System Topology & Satellite Access

Systems are designed around the topology and connectivity requirements of the sites that are interoperating. As illustrated in Figure 7.4, there are one-way (simplex) point-to-multipoint connections that distribute information to a group of users (e.g., multi-cast) or all users (broadcast). Further, two-way (full-duplex) communications can be fixed between just two locations (point-to-point) or with many locations and a central hub facility (star topology), or they can be switched as needed between or amongst many locations (mesh).

Typically VSAT systems are optimized in function, performance, and cost to fit one of these operating scenarios. A centralized hub facility serving many remotes allows one to minimize the complexity and costs of the remotes and place most of the intelligence/complexity/cost at the one hub location. Further, by using a large earth station antenna (8 to 14m) the satellite links from the remote to the hub can operate at very low power levels even when using 0.9m to 1.2m antennas at the remote sites. Thus the vast majority of the deployed VSATs today are of the interactive (star) type, because of their low remote costs.

VSATs

Mesh remote systems were originally designed for voice services in the early to mid-90's, as they were capable of switching satellite links between locations very quickly based on the demand (dialed telephone numbers) of users. This driving application, known as thin-route rural telephony, evolved into fairly low-priced terminals (below US $6K) that supported one or two phone connections per site and were deployed primarily into the Asia/Pacific and Latin America regions to provide voice infrastructure into rural and outlying areas.

Higher capacity, thick-route, mesh DAMA (demand-assigned-multiple-access) system were also developed in the mid-90s as they addressed the more traditional enterprise markets with voice and data capabilities. The first systems switched fixed rate SCPC carriers (typically 8Kbps for compressed voice) on and off between various sites based on demand. The thicker route systems allowed SCPC carriers of range of data rates (4.8K to 256K to 2 Mbps) based upon pre-determined connection requirements, AT or V.25bis commands or from the network management system. By 1997, IP routing functions were integrated into the DAMA VSAT terminal and ViaSat dubbed the capability DAMA-IP. The SCPC data connections and symmetric/asymmetric data rates were directly governed by routing tables based on destination IP address, source IP address, and type of application used (as determined by the source IP port address).

About the same time, COMSAT Labs had integrated frame relay functionality, with capacity and routing based on DLCIs (data link connection identifiers) which provided the needed address identification using permanent virtual circuits (PVCs) defined within their TDMA DAMA terminal and control system. Each PVC had a predetermined data capacity. COMSAT Labs soon followed with switched virtual circuit (SVC) capability and IP routing capability—the competitive race for IP-based intelligent VSATs was on in the late 90s.

The thick-route DAMA terminals were much more expensive then interactive VSATs and even SCPC VSATs. This is due to multiple modems being required for SCPC DAMA (one modem per active circuit), higher power RF required (because of a higher amplifier derating when operating in multiple carrier mode or the higher transmission rates associated with thick route TDMA) or the more complex and costly higher speed TDMA DAMA modems. Thick route mesh terminals in the mid-90s were around US $25K, decreasing to $15K by 2000 and are now in the $8K to $10K in larger quantities. This is due in part to COMSAT Labs being acquired by ViaSat.

Figure 7.5 illustrates the various capacity sharing techniques

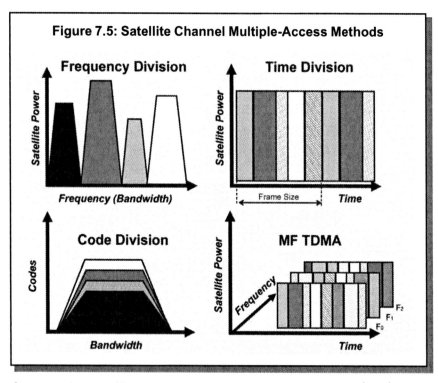

Figure 7.5: Satellite Channel Multiple-Access Methods

for accessing satellite bandwidth on fixed satellite services (FSS) satellites. The concept behind bandwidth sharing or bandwidth-on-demand is that connection capacity is not always used or needed unlike "nailed-up" SCPC circuits. By sharing capacity/bandwidth, more users can be served, capacity can be more efficiently and effectively allocated and the cost basis for per-unit capacity can be reduced.

A good example of resource sharing is a PBX in a larger company with hundreds of employees. While the PBX switch may have 200 telephone extensions for use by employees, there may only be 10 telephone line connections (trunk lines) to the central office. Statistically, not everyone within the company uses their phone at the same time, and even fewer are likely to make calls outside of the company. And in certain parts of the company, there may be only one phone for 5 or 10 employees. Thus, to minimize the cost of telephone service, 200-plus employees are served by 10 telephone lines connecting to the central office. For a manufacturing company, this 20-to-1 ratio might be perfectly acceptable, whereas for an enterprise having substantial public contact like an insurance or real-estate sales company, a ratio of 5:1 or 2:1 may be required.

Returning to our satellite access discussion, dividing up satellite capacity on a frequency basis, similar to AM or FM radio stations, is the basis behind frequency division multiple access (FDMA) as show in the top left illustration of Figure 7.5. This approach (also known as SCPC DAMA) provides circuit type connections between two locations. The bandwidth (capacity) and frequency of the system are assigned by a network controller based on available capacity and the determined need. Since satellite links can originate from anywhere in the satellite footprint the controller must determine the proper transmit power required in each direction to make the connection reliable. As shown in Table 7.1, SCPC DAMA is best used for circuit based services (voice, video, large file transfers) that have well-defined traffic patterns. Since each circuit requires its own modem, the number of simultaneous connections at a given site is limited by the number of available SCPC modems at that location. Thus as noted in Table 7.1, traffic patterns need to be understood in order to properly configure the DAMA terminal to ensure that capacity is available even during peak usage periods for a given location.

The most prevalent form of multiple-access is time division or TDMA. Here a single frequency is used for transmitting, with each network terminal allocated a small portion of time to transmit in. These transmit time slots have a minimum size (say 0.5 msec each) with 50 time slots per frame for example. The number of time slots allocated to a particular terminal during one frame time is predicated on the amount of capacity that is needed. Thus if a terminal transmitted for 10 consecutive time slots, then it would receive 20 percent of the capacity for that frame on our example system. The TDMA carrier transmit rate, known as the burst rate, is much higher (by a factor of 4 to 8 times) the typical data rate needed by the VSAT terminals. This is to ensure that efficient sharing of the burst carrier occurs. For instance, if the VSAT terminals require 32 Kbps of traffic, and you wanted to support 16 terminals per carrier then you would need a minimum burst rate of 512 Kbps (16 x 32K). Because of the limited capacity, one would never operate a single frequency TDMA network, rather frequency hoping multiple TDMA carriers would be used (MF TDMA). This scheme is illustrated in the lower right corner of Figure 7.6. Notice that a terminal may transmit on several different carriers during a single frame, as illustrated with the striped time slots. By allocating capacity over multiple TDMA carriers over various time slots, the sharing efficiency of the system is significantly increased as is the overall throughput.

The main reason TDMA DAMA and TDMA star technology is so widely used is the popularity of IP networking. The bursty nature of IP packets, even when providing voice-over-IP (VOIP) services, is a perfect fit for the burst transmission of TDMA. And as previously mentioned, star TDMA (or interactive TDMA) networks dominate the VSAT landscape because of their significantly lower remote terminal cost (below $2K) when compared to mesh TDMA (TDMA DAMA) at or above $10K per terminal for the same quantity.

The least often used access scheme is code division multiple access or CDMA. In this approach transmission occurs in both frequency and time coordinates with discreet channels based upon the use of different signaling codes. Each CDMA VSAT is assigned a unique code sequence which is modulated together with the user's digital data. The amount of bandwidth required on the satellite channel is a function of the number of codes, with typical spread factors of 100 to 500 and resulting bandwidth usage of 5 to 25 MHz. The main benefit of CDMA or spread spectrum is its significant resilience to transmission channel interference. Other benefits include no time constraints on when a terminal can transmit and each terminal has access to entire channel capacity. However, the large downside of requiring 5 MHz of satellite capacity (due to the spread factor) for a handful of terminals with 32 Kbps of traffic is significant. These systems are deployed only when the very small antennas (under 75cm) are needed, such as the low profile phase-array antenna used on commercial and business aircraft for delivering high-speed internet access.

Capacity Sharing—The Sums are Greater than the Whole

While there will always be a need for the traditional fixed-rate leased-line type services for certain applications and industries, many service providers are embracing the power of sharing network capacity across customers using bandwidth-on-demand VSAT systems.

End-users are warming up to the concept as well, especially as traditional terrestrial service offerings have migrated from leased line and T1 services, to frame relay, ATM and IP based fiber/wire services.

With traditional leased-line connections, the size of the data pipes (e.g., the data rate) is fixed, thereby requiring the data/voice capacity to be sized for the expected peak loading, to ensure the desired rate of service, throughput availability and quality of service (QoS). Once the data

VSATs

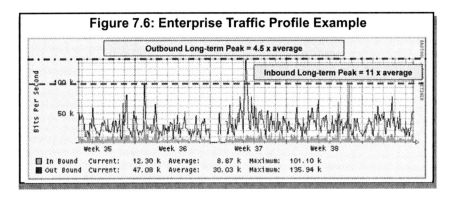

Figure 7.6: Enterprise Traffic Profile Example

capacity is selected and implemented using traditional SCPC services, the data rate does not change. If traffic requirements exceed the data rate, then the traffic is not accommodated and performance suffers. Likewise, if traffic requirements are considerably less, then capacity is unused. As a rule of thumb, when average traffic loads is 50 to 70 percent of available capacity, then fixed-rate services will provide the best-fit for user traffic.

However, SCPC end-users typically have dozens of sites, each with varying traffic capacity needs with traffic patterns that vacillate throughout a business day. Using a bandwidth-on-demand system, the unused capacity can be applied to locations that need additional capacity. This capacity allocation occurs automatically and very quickly, from under one second to a few seconds, depending on the VSAT system. As network capacity is shared amongst more locations, overall efficiency is improved even more, and the number of sites that experience capacity bottlenecks is further reduced.

To further illustrate our point let's take a real-world enterprise example, where there is a variety of data applications and voice/fax usage across a number of sites. Traffic analysis shows that over a five week period, the average (24 hour/day) capacity requirement is 12 Kbps (remote-to-hub) and 47 Kbps (hub-to-remote). However, peak traffic conditions where substantially higher, with hub-to-remote peaks 4.5 times the daily average and remote-to-hub 11 times average as shown in Figure 7.6. In evaluating various industry segments, it's not unusual to find intraday peaks of 20-to-50 times the daily average throughput for relatively brief periods of time (5 to 30 minute intervals). Furthermore, the busy-hour average is typically 2 to 4 times the traditionally measured 24-hour average, and considerably less than the daily peak average.

135

These types of traffic patterns can benefit significantly with bandwidth-on-demand (BOD) systems. The bottom-line benefit to end-users of a well-managed broadband-on-demand system—fewer performance bottlenecks, better service quality and lower overall service costs. For throughput sensitive applications like voice/VoIP, video conferencing and other real-time applications, the delivered quality-of-service is extremely important. Techniques for supporting real-time applications include a minimum throughput mechanism (e.g., committed information rate or CIR), traffic shaping, and prioritized queuing and transport (e.g., DiffServ, MPLS, etc.) thereby ensuring that essential traffic is transfered ahead of less-time-critical traffic.

Networking, Routing & Security

Prior to the mid-90s and the widespread adoption of personal computers and the Internet, computers and data networking were centralized, and communications protocols for the most part focused on mainframe computing or VAX type mini-computers. SNA, SDLC, HDLC and X.25 communication protocols dominated the landscape from 1985 to 1995. From a VSAT perspective, data was transported from the remote site to the central hub facility. The only special processing required of VSATs for data was handshake spoofing for protocols like X.25. Because of the approximately 350 msec round-trip delay over the satellite link, data reception acknowledgements between the date terminal equipment (DTE) at the remote location and the main data processor at the central location, would considerably slow-down data exchanges and remote application screens. To circumvent the acknowledgement delays, a protocol spoofing technique would be used such that the VSAT terminal would locally acknowledge the DTE and pass the data across the satellite channel to the central data processor. The hub VSAT equipment would also provide a local data acknowledge to the hub processor. The data transport across the satellite link was error protected and if any substantive data corruption was detected a retransmission request was sent to the appropriate local data terminal or processing equipment.

With the advent of personal computers and an effective means of interconnecting corporate networks via the Internet, an entire paradigm shift occurred in data networking. While the ATM cell-based frame relay services became the early transport favorite in the mid-90s, the Internet Protocol (IP) quickly assumed it dominance which continues through today.

Table 7.2 – Communications Services, Protocols and Routing

Category	Service Type	Addressing Protocols
Dedicated	Leased circuit or SCPC satellite	Permanent
Switched Circuit	ISDN (BRI/PRI) Video Conferencing Analog Telephone Digital voice	D-Channel RS-366 DTMF, Pulse, MF, R2 G.704, SS5, SS7
Packet	SNA, SDLC, HDLC X.25 Frame Relay TCP, UDP, FTP, HTTP VoIP	Native LAP-B DLCI IP IP/SIP

Source: MasterWorks Communications, 2007

As shown in Figure 7.3, the typical VSAT today has IP related functions integrated into the terminal. Routing mechanisms are especially important in mesh VSATs since their primary advantage is interconnecting VSAT sites based on traffic demands. Table 7.2 illustrates the various services, protocols and routing mechanisms that VSATs have supported over the years.

Specific to IP, functions that increase throughput speeds and transport efficiencies are of particular interest to manufactures and end-users alike. Given the prominence of TCP and the benefits of guaranteed delivery, vendors early-on had to resolve the slow-down in data throughput as a result of TCP acknowledgements similar that experienced in X.25 as previously discussed. Today, TCP/IP performance enhancement features have been built into many broadband VSAT terminals to significantly enhance throughput speeds.

Additional increases in throughput efficiencies and speeds have been realized by implementing IP header compression and even payload compression. Within the last few years, voice over IP (VoIP) has become increasingly popular and broadband VSAT vendors have integrated in real-time-protocol (RTP) compression (cRTP) to maximize throughput efficiencies. VoIP is very inefficient from a transport perspective because

The Satellite Technology Guide for the 21st Century

of the significant amount of overhead involved. By implementing cRTP techniques over the satellite link, voice calls can be supported within 8K to 9 Kbps as opposed to 18K to 22Kbps for non-processed VoIP packets.

Virtual private networking has become increasing popular in the past few years. A virtual private network (VPN) allows users to communicate and access information securely over the public Internet or other non-private IP-based networks. VPNs provide virtually seamless and secure connectivity that is akin to a private network, yet at a fraction of the cost by taking advantage of existing low-cost and widely available Internet connections.

VPNs provide secure end-end virtual connections or "tunnels" by encrypting the IP data payload (e.g., email, files, etc.). The encryption/decryption and other administrative functions (authentication and access control) occur at each end of the connection. These functions add additional overhead data requirements to the primary data connection, thereby requiring more packets to be transferred between end-points.

Some data encryption techniques like IPSec (IP Security) operate at Layer 3 (the network layer) and encrypt IP and TCP headers along with user data as a payload within the IPSec packets. Thus, the encrypted TCP header information is no longer "visible" to the TCP accelerator located with the broadband VSAT and throughput wanes considerably, even as much as 50 to 70 percent. One remedy for this type of encryption is to place TCP/FTP acceleration appliances to process the IP traffic prior to the encryption function. This acceleration appliance is therefore separate from the VSAT terminal.

Other alternatives in network security like Secure Sockets Layer (SSL) have gained in popularity and a growing number of hardware appliances and application suites are available. Because SSL works at Layer 4, it does not required specialized acceleration appliances.

More sophisticated networking capabilities are being implemented by VSAT vendors based on service provider requirements and end-user needs. For instance, when multiple tenants exist at a location served by a single VSAT, separate and distinct networks must be presented to each tenant. A typical example of this environment is an off-shore oil rig where multiple companies co-exist: the driller, the customer, the exploration service company, and the rig owner. Virtual LAN (VLAN) technology is used to provide a separate and secure LAN presence on physically separate Ethernet ports on the VSAT. Each VLAN is processed independently of each other at the remote VSAT, over the satellite link and at the VSAT hub facility. Ultimately, the individual network connections are presented to the destination companies' wide-area-network (WAN) connection and are terminated at their individual corporate facilities. In this way, true secure end-to-end data integrity is provided.

Deploying VSATs: Fixed or Mobile

Think of VSATs as enabling long distance wireless communications and you can almost visualize the many various locations around the world were they are deployed. Figure 7.7 provides a small sampling of VSATs at a wide variety of locations: fixed sites, trailers and transportable configurations, and mobile platforms like ships and trucks and even military vehicles.

Specialized antenna systems are the key to implementing satellite connections to transportable and mobile platforms. There are a number of antenna vendors who provide specialized tracking antennas with capable of operating on ships, barges, trucks, and even commercial and business aircraft. And recently, low-profile mechanically adjusted phase array antennas are being sold into the recreation vehicle (RV) and mobile home markets. In the near future low cost electronic steering flat-panel phase array antennas will be capable of transmitting and receiving broadband connections. These types of antennas are already in use in the military and for specialized deployments. Lower cost versions will revolutionize VSATs installation, operation and enable broader market penetration.

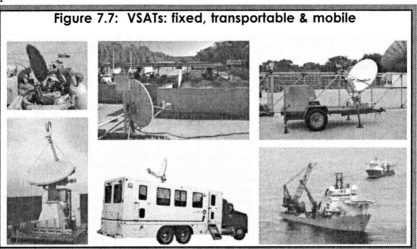

Figure 7.7: VSATs: fixed, transportable & mobile

The Satellite Technology Guide for the 21st Century

On-demand, real-time data streaming enables new levels of collaboration and expert decision making that substantially benefit enterprises.

Applications and the Real World

In the past ten years the worldwide volume of digital data traffic has been soaring – driven primarily by the Internet. As a result, data traffic is growing much, much faster than voice traffic. In fact, observers note that data traffic accounts for virtually all of the growth in commercial communications satellite traffic over the last five years.

The amount of data used in business, and the information required to stay competitive, is roughly doubling every six months. Access to the Internet, and a presence on the web, have become a necessity in today's corporate environment. Moving large amounts of data into, around, and out-of the enterprise network for national and multi-national corporations is becoming increasing more important and more difficult in a cost constrained, bandwidth-limited environment.

While fiber and other broadband technologies have become more widely deployed, no single communication technology can meet the capacity, geographical reach, and financial considerations that face many corporate IT managers. For economic, business and performance considerations a combination of network technologies is becoming more prevalent as the "best-fit" approach in corporate and government communications. In concept, this hybrid networking approach makes use of the "best" features of fiber, wireless, wired-facilities and satellite technologies, while minimizing the "rough spots" that might occur at the seams where these technologies fit together. Therein lies the challenges facing equipment manufacturers, integrators, service providers and users of hybrid networks. Fortunately, these issues are manageable and in fact, the majority of satellite networks deployed today are hybrid satellite/terrestrial networks.

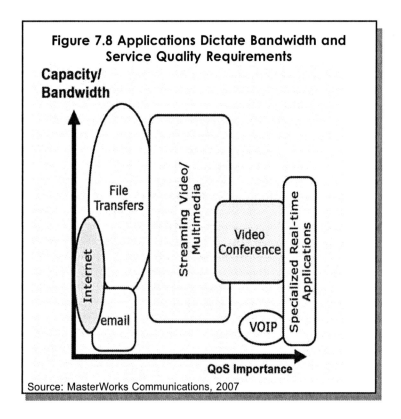

Figure 7.8 Applications Dictate Bandwidth and Service Quality Requirements

Source: MasterWorks Communications, 2007

Figure 7.8 illustrates the many applications that face IT managers today—and there are many more; but fortunately, most, if not all, are supported in an IP networking environment. These applications determine the capacity and QoS requirements of the network. If Internet access is all that desired, then service quality is not very important and service bandwidth needs are moderate and network implementations are less challenging. However, voice, video conferencing and specialized real-time applications require a high level of service quality (low latency, consistent throughput and reliable connections) and can be particularly challenging to deploy over wide geographical areas or capacity diverse networks.

A Case Study: How VSATs are Transforming Business

As a case example of a successful hybrid network deployment with demanding real-time broadband applications, we'll turn to the energy service sector. Over the past 5 to 10 years there has been a definite long-term uptrend in increasing data rates for the oil and gas sector. This is a result of the greater use of networked IT and the driving economics of getting well logging data back to base for almost real-time analysis by specially trained experts.

The energy exploration and production industry is a very economically competitive business—lose a few days of production or miss the critical decision point during a drilling project can mean a loss of millions or hundreds of millions in revenue.

Travel expenses and lost employee productivity due to travel time are significant to many companies around the world. But to a U.S. based Fortune-100 oil services company, lost productivity of critical expert resources was the key motivation for finding a better way to do business. Before the company deployed its satellite-based broadband wide area network (WAN) across the Americas, West Africa and Europe, its experts went to where the geological data was – often spending half of their time in non-productive travel, getting to exploration sites at sea or in hard-to-reach land locations many hours or even days after departure. Once on site, they would analyze large amounts of data captured in real-time using specialized computing applications, and make drilling or process recommendations on the spot, yielding immediate improved results.

This highly specialized service has brought in big money for the company, but the number of jobs that each expert can support limits company revenues—that is until now. By integrating broadband VSAT capacity with smart IP routing capability and their existing corporate WAN backbone this company, along with many others, are successfully deploying broadband VSAT-based WANs that bring together their LANs, which are thousands of miles apart, at speeds of 1 to 2 Mbps or more.

Since they've deployed their broadband WAN infrastructure, the high-volume, real-time, well-site data comes to their experts, who are regionally located in offices and who can now support several operations at multiple sites concurrently. The results—increased productivity, larger service revenues and happier employees. And when local conditions warrant, expert on-call help is just a videoconference or a telephone call away for the drill crews in the field. And for crews that remain on-site for offshore rig or shipboard operations, the new network gets used off-hours as well—for calls home, email or videoconferncing. Thus, the company has significantly altered the model they use to deploy their geological services and plan on expanding services into new markets.

VSATs

Table 7.3: High-profile Corporate VSATs Users

Industry Segment	Corporate VSAT User
Food` Restaurants	Taco Bell, KFC, Carl's Jr. Cracker Barrel, Domino's Pizza Jack-in-the-Box, Sonic, Wendy's, Long John Silver, McDonald"s Cracker Barrel,Outback, Uno, Perkins, TGI Fridays, Boston Markets
Grocery Distribution	Safeway's, Smith's Food & Drug, Weiss, Traders Joe's, Albertsons
Medical Drug	Pfizer, Thrifty, Rite Ad, Shoppers Drug Mart, Kerr Drugs, Blue Cross Laboratories (India), Pharma Plus
Car/Truck Rental	Avis, Hertz, Enterprise, National, Ryder
Automotive	Goodyear, GM, Ford, Volkswagen, Saab, Peugeot, BMW, Nissan O'Reilly Autoparts, Pep Boys, Volvo, Daimler-Chrysler, Mercedes Porsche, NAPA Auto Parts, Jiffy Lube
Energy	BP, Mobile Oil, Texaco, Stato I, TotalFina Bf, Schlumberger Halliburton, Amoco, Dhevron, Arco, Unocal, Nabor's Drilling, Shell Sinclair Oil, Suncor, Texaco, Bharat Petroleum, Sunoco, Imperial Oil
Banks	World Bank, Abbey National, Chase, Asso. BankCorp, National Bank of Romania, Nova Bank, Citibank, First National, National Bank of Poland
Financial Services	American Express, Morgan, Stanley Dean Witter, Smith&Barney Edward D Jones, Master Card International, Countrywide Home Loans
Stock Exchanges	Shanghai, Bombay, Delhi
News & Financial Information	UPL AP, Bucharest Media, Reuters, Bloomberg, Malaysian News Agency (Bernama)
Manufactured Goods	Coca Cola, Colgate Palmolive, Nestle India, Samsung
Print Media	USA Today, Gannett, Walden Books
Hotels	Quality Suites, Best Western, Choice Hotels, Holiday Inn
Retail	Sears, Williams Senoma, Big 5 Sporting Goods, Michaels Arts Crafts Sherwin Williams
Clothing	Cargil, Kohl's, Rack Room Shoes, Ross, Smart & Final, Foot Locker TJ Max, Casual Male
Entertainment	Blokbuster, Hollywood Video, Warner Bros, Brunswick Bowling Musicland, Regal CineMedia, Video Jukebox, Video Venue
Airlines	L'Air Liquide, United Airlines, Jet Blue, China Eastern Airlines
Military	US Air Force, Navy, Army, Marine Corps, Coast Guard
Government Services/Transportation	US Post Office, UK Post Office/Energis, Department of Posts (India) Mailboxes, Etc., Public Storage, Union Paciific Railroad, Penn Traffic Namibian Railways, ARINC, American Farm Bureau, John Deere & Company, Federal Express, Yellow Freigths, Australian State Railways
Lottery	Hungarian State Loterry (SZRT). UK, South Africa (Uthingo), Spain (STL), China (Welfare)

Source: Comsys VSAT Report 2007

Many companies and governments around the world have deployed and operate VSAT-based data and voice communications networks. Table 7.3 highlights a number of recognizable companies in a variety of industries, that are dependent upon VSATs within their corporate network infrastructure.

Services or Equipment—How to Choose

On the surface it may seem like a tough question to determine if a company should purchase VSAT-based services from a service provider or implement your own service. As we have already read, many large corporations rolled out there own systems in the 1985 to 1995 time frame. They learned pretty quickly that it takes a huge investment in resources, time and commitment. Most made this decision because the number of qualified service providers was limited, especially during the early years. However, in the past ten years, by far the majority of VSATs services are outsourced to VSAT service providers.

What does it take to roll-out a VSAT-based service? VSAT equipment must be selected and purchased, vendor training conducted, the VSAT network must be engineered and then integrated with the existing telecoms, computing network and business operations, satellite space segment must be purchased, licensing obtained, hub earth station installed and a small pilot system rolled out and tested. Additionally, network operators must be trained along with field support and help desk personnel. This is followed with a rollout of the remaining VSATs to all of the business locations. In short, a whole lot expertise and activities are required to successfully launch and operate a broadband VSAT system.

As the third and fourth generation bandwidth-on-demand (BOD) VSAT systems rollout, service providers and their corporate customers still seem to be mystified by the available technologies, the accompanying service offerings, and how to select amongst the alternative system offerings.

In providing a service offering, the most fundamental concern is quantifying the end-user's requirements. While there are many considerations to make, the most important can be distilled down to these three:

- How much capacity is needed?
- How important is quality-of-service (QoS)?
- How much is the end-user willing to pay?

These three characteristics are tightly interrelated; the better the QoS and higher the capacity, the more the service will cost. As previously illustrated in Figure 7.8, the end-user's applications will determine the capacity and QoS requirements. If Internet access is all that desired, then service quality is not very important and service bandwidth needs are moderate. However, voice, video conferencing and specialized real-time applications require a high level of service quality (low latency, consistent throughput and reliable connections). And for service providers, customer requirements vary by industry segment (e.g., financial, oil

Considerations in Selecting an Equipment Vendor

& gas, disaster recovery, construction, government, retail, etc.) making network design and operation a carefully considered endeavor.

For service providers, selecting a VSAT equipment vendor is a very important decision as the economic and resource investment is significant and will last for years. Because of the growing complexities of implementing and operating a network, the decision should be viewed more as a partnership than a buyer-vendor relationship. You'll likely be doing business with this manufacturer for 3 to 5 years minimum.

Therefore, select a broadband VSAT system platform and an equipment vendor that:

- Is capable of providing sufficient capacity for both in-bound (remote-to-hub) and out-bound (hub-to-remote) directions and one that can scale easily to accommodate more capacity, more terminals, or more end-user traffic per terminal;

- Has a successful track record of delivering their products into similar application needs; if a vendor specializes in delivering Internet access to schools, for example, the product may not perform well for lots of voice traffic and large file transfers;

- Integrates quality-of-service (QoS) mechanisms/tools that support traffic prioritization, integrated protocol support, routing, and capacity load balancing;

- Has attractive economics; evaluate cost of infrastructure across expected client base (capital costs), remote site implementation costs and operational costs (bandwidth efficiency, load-balancing techniques/capacity management tools and maintenance);

- Has very good network planning, monitoring and service level agreement (SLA) management tools; this is especially important if you offer SLAs to your customers;

- Has good economic and business stability, coupled with a solid reputation, especially in post-sales and technical support;

- Has an established product line offering—remember one size does not fit all;
- Is forward looking, having multi-year product and business plans.

Selecting a system strictly based on network efficiency can be problematic as network efficiency is difficult to quantify in a global sense. Efficiencies can be measured in different ways, and efficiency can vary, even on a single system, depending on traffic levels, profiles and applications. In effect, if a system is to support a variety of applications and customers in different market segments, system throughput and efficiency measurements will vary from day to day.

Factors that affect bandwidth, throughput efficiency, and performance include:

- TDMA frame size (fixed or flexible)
- Modulation (QPSK, 8PSK, etc.)
- Forward error correction (FEC) coding types (turbo product coding vs. LDPC vs. Reed-Solomon/Viterbi (RSV) and operating mode: static (fixed) or adaptive (system automatically changes FEC rate/type depending on operating conditions)
- Channel access methodology (slotted aloha, reservation: fixed or dynamic, etc.)
- Antenna sizes (remote and hub)
- IP packet transport methodology and data packet sizes
- IP applications (e.g., protocols: TCP, RTP, FTP, UDP, etc; real-time vs. non-real-time, etc.)
- Quality of Service (QoS) provisioning
- IP protocol performance enhancements (TCP acceleration, RTP compression, etc.)

Delivered service offerings—not all are equal

From an end-users' perspective outsourcing your telecoms, especially for service to locations that are best served by satellite, is generally a very wise decision. While there are hundreds of VSAT-based service providers around the world, there are likely only a handful that serve your particular locale or industry segment. And selecting a service provider can still be challenging as there are many ways of packaging and delivering broadband-on-demand services, especially to demanding locations like off-shore oil platforms, remote construction or mining sites, ships, etc.

What is important to you as an end-user is that you determine how likely your perspective service provider is able to successfully deliver a quality network service that meets your technical, operational, and business requirements. Look for a service provider that can deliver:

- Technical depth, expertise and proven track record for thoroughly integrating their service solution into client corporate networks and infrastructure—especially for business operations similar to yours;

- Networking expertise that provides seamless end-to-end operation for even your most demanding application needs

- The capacity you require with a high Quality of Service (QoS) that is backed by a Service Level Agreement (SLA)—usually only service providers that are confident in their ability to deliver a quality of service will agree to SLAs;

- Reference customers who have similar networking requirements to yours—contact them and understand first hand their experiences (good and bad) with this service provider;

- A complete service price with a clear understanding of the service being provided (including an SLA) and that there are no hidden charges or required future "upgrades" for additional charges.

With respect to the technical details ensure that you understand the following about the offered service:

- What is the delivered capacity per site? (maximum and minimum; is there a guaranteed throughput, e.g., CIR?)

- How often is the capacity adjusted? (seconds, minutes, hours, etc.)

- Is the capacity shared with other end-users? If so, using what oversubscription rate?

- How is quality-of-service (QoS) implemented, managed and audited? (per site and across your network of sites)

Comparing the technical merits of each of the systems available in the marketplace is primarily of interest to service providers. From an end-users perspective much greater importance is typically given to the quality of the service that the service provider delivers.

The latest generation of broadband VSAT systems is based on complex technology, which requires sophisticated network design, system integration and operational capabilities by network operators/service providers. Unfortunately for end-users, not all service providers have the ability to meet these challenges. In short, a service provider may have the very best VSAT system, but service delivery may not be as expected because the provider:

- may not understand the complexities of broadband VSAT technology network design, implementation and operations;
- has not integrated or installed the service properly;
- has too many users on the network resulting in significant congestion and poor performance;
- may not understand the intricacies of IP networking over satellite, quality of service (QoS) or delivering a consistent and reliable grade-of-service from end-to-end.

When you decide to look for a service provider, shop around. But resist the temptation of purchasing services from the lowest bidder, especially if they are considerably less than the others. The better service providers may cost more, but you will also have peace of mind when you call them for support and they answer the phone.

Emerging Trends In VSATs

Like most high-tech industries the only constant in the VSAT industry is change. Since the industry was born some 25 years ago, there have been tremendous changes across the years. Technology has made leaps and bounds, business practices and service delivery have improved—becoming more efficient, more responsive and more customer centric. Dozens of equipment manufacturers, service providers, integrators and even satellite operators have come and gone with consolidation occurring almost everywhere. Yet this is an exciting industry that has overall stability, growth, and heritage—this industry is constantly moving forward meeting each emerging challenge, be it technical, economic or business.

Standards and their Role in Shaping the Future

All broadband VSAT systems that operate with a centralized hub facility utilize separate inbound (remote-to-hub) and outbound (hub-to-remote) channels. Various VSAT platform manufacturers take different approaches in implementing their systems. Proprietary approaches emphasize operational efficiencies, performance and innovation, while open-standards based systems leverage existing technology (e.g., DVB-S) for lower implementation costs or enable multiple vendor sourcing (e.g., DVB-RCS) to achieve economies of scale and vendor independence. Overall, the industry is in violent agreement that end-users desire low-cost, high-performance VSATs with product longevity but the clear topic of debate is how to achieve this goal.

To date proprietary systems have dominated the landscape and account for 90+ percent of currently fielded systems as shown in Figure 7.9 below. The DVB return channel standard (RCS) was adopted in 2000 by ETSI (European Telecommunications Standards Institute, EN 301 790), with a hand-full of manufacturers developing and fielding DVB-

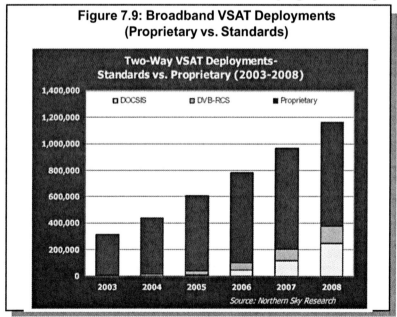

Figure 7.9: Broadband VSAT Deployments (Proprietary vs. Standards)

RCS systems. Even with a number of satellite operators like Eutelsat and New Skies adopting the technology and implementing service offerings, fewer than 20,000 terminals have been delivered by 2006 by all vendors. Under the guidance of ESA's independent test initiative, SatLabs, a number of systems vendors conducted preliminary testing to demonstrate interoperability in mid-2003. The formal SatLabs' qualification and certification program was launched in April 2005. And in the past several years over a half-dozen vendors have received DVB-RCS compliance certification.

One significant issue with the DVB-RCS standard is that it specifies only the satellite channel interface or "air standard" and how the hub functions in terms of bandwidth management, control and message handling. The standard does not address considerations such as IP acceleration, routing functions, quality-of-service implementations and other service orientated functions that occur within the VSAT terminal and hub baseband equipment. Overall these can have a significant impact on delivered service quality and speed.

While ViaSat provides a DVB-RCS compliant VSAT with their LinkStar system, they have also embarked on a new return channel approach based on the DOCSIS® (Data Over Cable Service Interface Specification) networking standard, used in cable delivery systems throughout North America. This rather innovative approach is at odds with the DVB-RCS standard.

Part of the motivation for using the DOCSIS standard is to take advantage of the large back-office tool suite available for operating and administrating huge subscriber populations—certainly a key concern for any success minded service provider. ViaSat has already teamed up with Intelsat to deliver over 10,000 DOCSIS Ku-Band SurfBeam terminals for two-way Internet services in the Middle East and it's delivering this technology for the Wild Blue Ka-band service which launched in 2005. They have also teamed with Eutelsat, Telesat, SES Americom and operators in Asia to deliver SurfBeam as well. By early 2006 ViaSat had delivered 100,000 Ka-band DOCSIS-based VSATs to address the consumer market and by the end of 2007 ViaSat has delivered more than 300,000 of their SurfBeam terminals to consumer and SME customers.

Not to be left behind in the standards issue, Hughes Network Systems worked closely with the TIA (Telecommunication Industry Association) in early 2004 to have its existing DirecWay air interface technology published as a standard under the Internet Protocol over Satellite

IPoS title. ETSI ratified the IPoS standard in early 2005. While Hughes is delivery royalty-free technical documentation and fee-based technical support and technology licensing, it's completely unclear why other vendors would consider manufacturing standards compliant hardware when the design is relatively dated and provides less capacity and scalability than either DVB-RCS or DOCSIS. To their credit both of those standards were designed for very large terminal populations—one million plus. HNS claims to have over 500,000 IPoS compliant terminals fielded, with shipments in 2004 topping 180,000.

The debate over DVB-RS and DOCSIS for satellite return channel technology will continue for the next several years until there's a clear production winner—likely not to be declared until 250K units or even 750K units have been shipped. Once these bigger volumes are achieved, the US $1,000 barrier will finally be broken and the new target will be below US $500 for Internet access delivery.

Technology Trends and Advances

Fortunately, the successful equipment providers continue to listen to their customers and market needs as they strive to maintain market leadership. System reliability, value and performance are at the top of most customers' must-have list.

We are likely to see continued consolidation amongst equipment manufacturers, especially the smaller ones. Equipment vendors that have staying power are keen on differentiating themselves amongst their peers and continue to capture more market share—even if their target are more niche in nature. For example, one new vendor to the industry is just beginning to rolling out their product offering—a single-carrier TDMA system that provides native IP transport with mesh connectivity and operates without a centralized controller.

Over the near future you will find mainstream broadband VSAT vendors roll out value-add capabilities to their existing product lines. iDirect for example, has just rolled out a mesh capability to their Infiniti product line. There is definitely market demand for hybrid star-mesh services, where most of the network traffic flows from remote locations to the hub-gateway, but some traffic, especially voice, needs to occur between remotes and regional offices. With somewhat limited capability at first, iDirect will soon follow with full featured mesh capabilities.

Another example of technology advances will be the increase of

The Satellite Technology Guide for the 21st Century

link performance and reliability using adaptive modulation and forward error correction (FEC) techniques. The idea behind these techniques is to operate with the highest efficiencies possible for the majority of the active terminals, such as 8-PSK modulation and rate 7/8 turbo product codes for example, and dynamically "downshift" for remote sites that are experiencing rain or other transmission impairments (e.g., QPSK, rate ½ LDPC). This downshift would only occur when communicating with a disadvantaged terminal, which could be for a small percentage of time each system frame. Because throughput capacity is sacrificed for reliable connectivity, terminal throughput decreases, but only while this impaired operation is occurring for this location.

Some of this capability already exists in the ViaSat SurfBeam system. iDirect has implemented a limited version of this capability into it's product line. Comtech EF Data has even implemented a dynamic

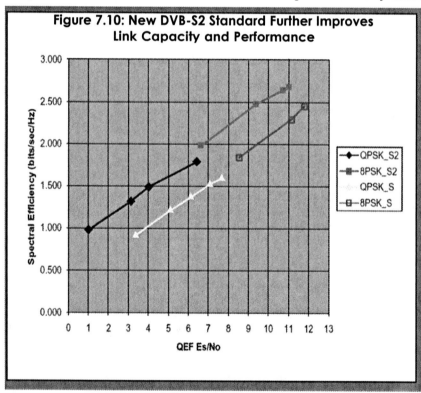

Figure 7.10: New DVB-S2 Standard Further Improves Link Capacity and Performance

Source: Radyne Comstream White Paper on DVB-S2 and the Radyne Comstream DM 240.

VSATs

coding capability for its SCPC modems that is under network control to increase reliability at edge of footprint locations for maritime applications. As the vessel sails into a less advantaged area of the satellite footprint, the modem controller changes the coding rate while maintaining the same symbol rate. This ensures that the RF spectrum is maintained in both power and bandwidth. Of course the throughput is reduced accordingly, but the link continues to operate at the desired availability level.

These advanced coding and modulation techniques will continue to be further improved upon and become available across multiple vendors in the months and years ahead.

Another technique to improve utilization efficiency of bandwidth on the satellite is to operate two SCPC carriers on the same frequency, at the same time. Impossible you say? Yes unless for the magic of digital signal processing and an old idea that been around for awhile—adaptive signal cancellation. Two companies have this technology implemented— ViaSat with their paired carrier multiple access (PCMA) technology and Comtech EF Data with their carrier-in-carrier® technology licensed from Applied Technology who names it DoubleTalk™. Both ViaSat and Applied Technology claim patent ownership and both companies use adaptive signal cancellation to allow the transmit and receive carriers of a full-duplex link to occupy the same transponder frequency and bandwidth. Figure 7.11 below illustrates the "before" and "after" image of the satellite spectrum. Because the each transmitting site "knows" what it is transmitting, that site can "subtract out" or cancel it's own transmitted signal to "see" or receive the transmitted signal from the far end.

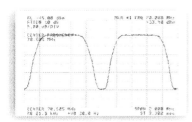
Figure 1:
Without DoubleTalk Carrier-in-Carrier

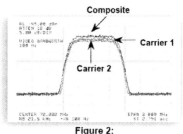
Figure 2:
With DoubleTalk Carrier-in-Carrier

Figure 7.11 Satellite Spectrum Without and With DoubleTalk Carrier-in-Carrier

This technique works best for links that are heavily bandwidth limited, meaning they use more bandwidth than power on the satellite. The concept is to have both carriers occupy the same bandwidth and each use slightly more (0.2 to 0.4 dB) transmit power. So to achieve maximum savings, the full-duplex carriers would operate with equal transmission and coding rates and identical modulation. Each would have an occupied bandwidth slightly greater than twice their allocated power. So when you combine the two carriers on top of each other, their composite bandwidth is the same as just one carrier and their composite power is slightly greater than their stand-alone transmit power (for the same link quality).

The advantage of this technology is best realized for new installations, where larger antennas may be used to "make" the link more power efficient and thereby create a bandwidth limited condition. Likewise when evaluating existing links to determine the transponder cost savings, a lower modulation and code rate is typically used to "spread out" the individual carriers thereby making them bandwidth limited and reducing their power requirements somewhat. While the theoretical savings of this technology is 50 percent, the typical savings is 20 to 30 percent of existing transponder costs. With this level of savings it's possible to "pay-off" the cost of these special modems well within the first year of ownership.

Other trends in technology centered around the product implementation. Some VSAT vendors are moving towards an open-platform hardware approach for their remote VSATs. This enables customized applications and capabilities to be integrated and "hosted" within the VSAT terminal. Such functions could include content caching, data pre-fetch processing, traffic shaping, specialized data compression, firewalls, virus and hacker protection and the like. Each of these add-ons enhance the product for a particular applications use.

Terminal pricing will continue to fall somewhat, especially as production volumes increase, but there will always be room for feature rich VSATs for higher-end enterprise solutions that will command, and get, top price.

Services: How Providers Will Adapt to a Changing World

Given the complexities of today's telecom environment there are many challenges and nuisances of delivering end-to-end communications services. The many successful service providers worldwide are the true unsung heroes of the VSAT industry. Their success requires them to be very good at a wide-range of disciplines; satellite communications, computers & technology, data networking, telephony & video, system integration, equipment installation and even civil engineering at times, operations, and customer service. And service providers operating in mobile, off-shore or hostile environments (oil rigs, ships/vessels, energy exploration, SNG, government/military) face even greater challenges. Service providers provide an important and essential role in bridging the very big gap between end-users, equipment manufacturers and satellite operators. They are quite literally, the glue that holds the industry together.

Successful service providers will continue to meet the challenges of their markets and customer needs. Competitiveness is everywhere, so the most successful will continue to demonstrate business leadership, technological awareness, quality service delivery and customer satisfaction. In the end, service delivery and reliability have more staying power and value to customers than having the lowest price or the latest technology.

Author's Note:

This chapter is dedicated to Dr. Irwin Mark Jacobs, a true pioneer in satellite and wireless communications, a father of many companies and a mentor to many men and women. My successful twenty-five plus year career in satellite communications has been heavily influenced by Dr. Jacobs and the many talented men and woman who gathered together to change the world at a company once called Linkabit in San Diego. I was fortunate enough to start my career at Linkabit and as a result I have a front row seat and a hand in the development of many of the systems and the history written about in this chapter while I was at Linkabit and at two other Linkabit spin-offs: ComStream and ViaSat.

The Satellite Technology Guide for the 21st Century

      I learned a great deal from this man and those that have followed his ways and his pioneering spirit. Thank you Irwin Jacobs—you have made the world a much better place.

--John M. Puetz.

CHAPTER 8
Satellites and the Internet

by DC Palter

The advent of the internet for consumer entertainment and business communications brought new opportunities for satellite providers and new challenges as well. Data, voice, video and other services now all ride over the internet and satellite operators and equipment manufacturers have focused their attention on providing IP connectivity.

Because of the high cost of building, launching, and operating satellites compared to terrestrial solutions, satellites are rarely cost effective in locations where terrestrial alternatives are plentiful. However, satellites do possess the unique advantage of ubiquitous coverage anywhere in the world, providing network access to rural, remote, and rugged areas where it is not cost effective to lay fiber. Further, satellite networks can be installed quickly, making satellites ideal for emergency situations. And new broadband satellites and low-cost standardized equipment offer the promise of wide-scale deployment in the near future.

IP Networks

Although almost all computer networks use the same underlying internet technology, the needs of ISPs, consumers, and businesses vary considerably, and the satellite networks to support them are quite different.

ISPs

Internet service providers (ISPs) use satellites to extend the internet backbone to locations where fiber isn't available or where local

The Satellite Technology Guide for the 21st Century

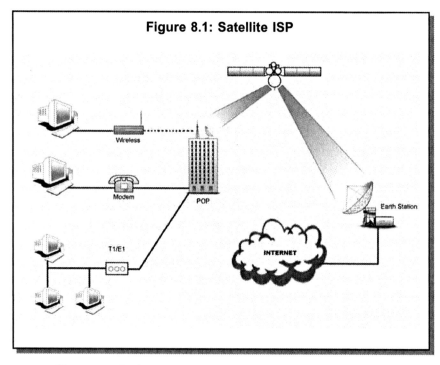

Figure 8.1: Satellite ISP

monopolies or regulations make terrestrial solutions impractical. A satellite earth station at a local point-of-presence (POP) backhauls internet traffic through a teleport, usually in a major Western city, where it connects to the internet backbone.

Local customers connect to the POP through dial-up modem, DSL, T1/E1, wireless or any other networking technology. In most cases, the end user is unaware that behind the scenes, the ISP service is provided by satellite. Most ISP networks use dedicated SCPC links, but some regional ISPs employ star networks to share bandwidth between multiple POPs (See Figure 8.1).

Satellite backhaul is especially prevalent in Africa and the Middle East as well as parts of Asia and the United States and Europe. A number of satellite providers specialize in offering backhaul service to local ISPs, most prominently Intelsat and SES Global.

Consumer Networks

Satellite-based ISPs allow consumers to connect to the internet directly over satellite. These services use an inexpensive VSAT system, with a small dish easily mounted on a roof or balcony. Within the United States, the three major consumer satellite ISPs are WildBlue, HughesNet, and StarBand.

First generation services, which predated wide-scale availability of DSL and cable modems, were mostly one-way systems, offering a satellite downlink but requiring a telephone line for the return channel. However, consumers were not fond of the added cost and inconvenience of the hybrid architecture, and in addition, a special card had to be installed inside the computer along with associated software. This was inconvenient for non-experts, while limiting power users to a single, Windows-based computer.

Second generation systems solved these problems with a two-way satellite service connected through a small set-top box (See Figure 8.2). Data travels from the consumer over the satellite to the ISP's earth station where it reaches the internet backbone, providing easy, inexpensive internet access anywhere within the satellite's beam.

Still, consumer satellite broadband is a challenging business. Consumers expect prices comparable to terrestrial options. Even in remote communities, users are unwilling to pay much more than $50 per month. However, unlike terrestrial services, where the cost of bandwidth itself is negligible, satellite ISPs have to overcome the high price of bandwidth while subsidizing the expense of specialized equipment and professional installation.

For example, if it costs the ISP $2 million per year to lease a 45 Mbps transponder, the bandwidth cost alone for a 1.5 Mbps downlink and 500 Kbps return channel is $7400/month. Even 100 Kbps shared between uplink and downlink, equivalent to a dial-up modem, costs $370/month in bandwidth.

In order to keep prices low for consumers, it is therefore necessary to oversubscribe the bandwidth, but this is a delicate balancing act. Putting too many users on the link leads to oversaturation, slow performance, and customer dissatisfaction, while too few users makes profitability impossible. With usage fluctuating over the course of the day, it is difficult to find a balance that uses all of the expensive bandwidth during off-peak hours without suffering overly poor performance during peak loads. And with small businesses, cybercafés, and even ISPs freeriding on the inexpensive consumer service while power users chew up bandwidth sharing music and video files, the business becomes even more difficult to sustain.

The Satellite Technology Guide for the 21st Century

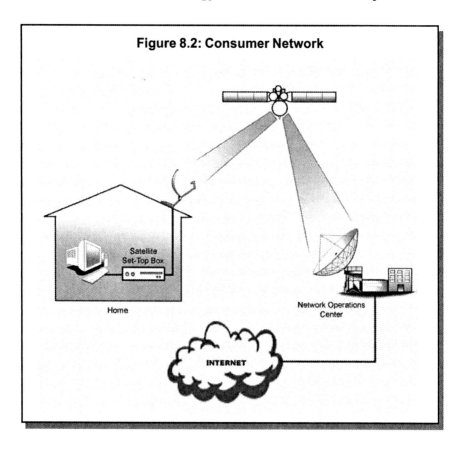

Figure 8.2: Consumer Network

However, a few solutions help to mitigate these challenges:

- Though unpopular, a fair-use policy is essential to limit bandwidth hogs.

- Offering both business and consumer services provides a mixture of traffic during daytime and evenings hours and adds higher margin customers to help offset operating expenses.

- Providing the service across multiple time zones flattens peak loads.

- During late-night hours when bandwidth utilization is low, multicasting cache contents, movies and files to reduce peak traffic loads or generate marginal revenue.

Business Networks

Businesses need to connect remote offices to a central corporate location. For the smallest sites, especially telecommuters or local offices with minimal data traffic, a consumer-oriented service is usually sufficient and provides the most cost-effective solution. The local office is then connected to headquarters over the Internet using a VPN tunnel for security (See Figure 8.3). Larger enterprises with many remote sites usually require a dedicated VSAT network. An earth station is constructed at corporate headquarters and VSAT terminals are installed at each local office. Users access e-mail, Citrix, database, and other applications over the satellite (See Figure 8.4). This is no different from traditional frame relay and other corporate WAN architectures, except that the WAN link runs over satellite. A star topology shares the bandwidth across multiple sites, making it the most cost-effective solution for most situations.

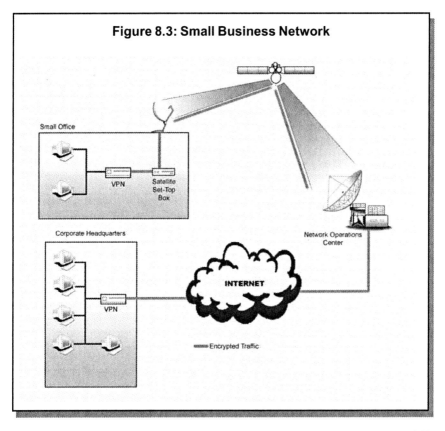

Figure 8.3: Small Business Network

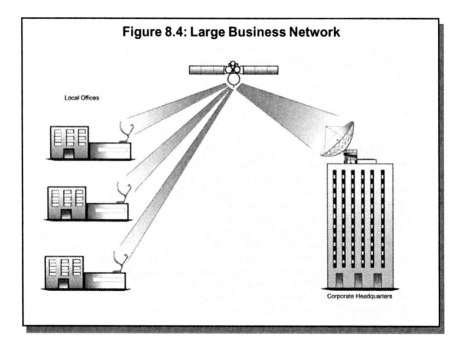

Figure 8.4: Large Business Network

Medium-sized enterprises generally have the same needs as bigger companies, but with fewer sites, the cost of constructing and operating a dedicated network may not be feasible. Instead, many teleport operators offer a shared network service with guaranteed minimum bandwidths and the ability to burst to higher speeds when bandwidth is available. In most cases, the company runs a dedicated line from corporate headquarters to the teleport, with an end-to-end VPN to provide security (See Figure 8.5).

Hybrid Networks

Satellite has traditionally been used to connect local terrestrial networks. However, recent advances in wireless technology are now ushering in the age of hybrid networks, combining satellites with a wireless local network.

Wi-Fi has gained considerable popularity within the past few years to provide access throughout homes, small businesses, hotels, and cybercafés. A small, inexpensive wireless router obviates the need for cumbersome cabling, providing mobility and flexibility while reducing costs. Combining a Wi-Fi router with a VSAT terminal makes possible simple, inex-

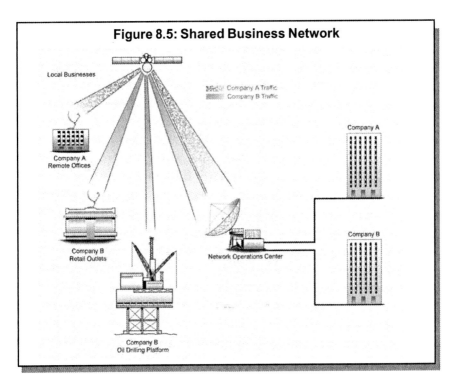

Figure 8.5: Shared Business Network

pensive internet connectivity anywhere in the world with minimal set-up time and expense (See Figure 8.6). Consequently, Wi-Fi is increasingly being built into consumer satellite routers and low-cost VSAT terminals.

While Wi-Fi is ideal for homes and the local coffee shop, its range is limited to less than 1000 ft under the best conditions and around 100 ft inside typical buildings. In contrast, WiMax, a new wireless technology, can cover an entire town with a single tower. WiMax has a range of about one to three miles in the city and 10 miles or more in rural areas. Initial rollout began in 2006.

Satellite and WiMax are an ideal combination for offering internet access in remote villages. With the installation of only a VSAT terminal and a single tower, an ISP can offer broadband service to an entire town. However, WiMax is a mixed blessing for the satellite industry. Although it opens new opportunities for satellite backhaul services, it seems likely to reduce the need for satellite-based consumer broadband offerings such as HughesNet and WildBlue.

The Satellite Technology Guide for the 21st Century

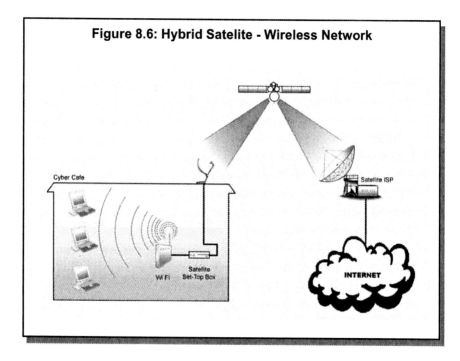

Figure 8.6: Hybrid Satelite - Wireless Network

Satellite ISPs

There are hundreds of satellite ISPs around the world, but most are not well known outside their local region. A few of the larger providers are HughesNet, Starband, and Inmarsat.

HughesNet

Hughes Network Systems' (HNS) HughesNet (formerly DirecWay and DirecPC) is one of the original satellite ISP services, and at more than 1,000,000 units, by far the largest. HNS offers a variety of packages for consumers, small businesses, enterprises, and government organizations, and through subsidiaries and partnerships, is available across most of the world. A typical single-user package in the U.S., with downloads speeds of 700 Kbps costs US $59 per month with a US $599 set-up charge for hardware and installation.

StarBand

StarBand, introduced by Gilat in the U.S. in 2000, primarily focuses on consumers, telecommuters, and small offices. A typical single-user package with a one year service contract costs US $69 per month for 500 Kbps download speeds and US $399 for equipment. Gilat offers the same system to national and regional operators throughout most of the developing world.

Inmarsat

Initially formed as a treaty organization to provide global communications for the maritime community, Inmarsat is the premier supplier of mobile phone, fax and data communications by satellite, with more than 287,000 ships, vehicles, aircraft and other mobile users. Inmarsat is used primarily for telephony, but some Inmarsat terminals can provide internet access at up to 128 Kbps. At the end of 2005, Inmarsat launched its Broadband Global Area Network (BGAN) service, offering speeds up to 492 Kbps.

Inmarsat charges by the minute for connectivity and the costs can be steep. It is therefore used for occasional e-mail, telephony, and critical file transfers by mobile users rather than broadband access.

GEO, MEO, and LEO Satellites

Traditional communications satellites orbit at a height of 22,300 miles (36,000 km) above the earth. At this precise location, the satellite orbit takes 24 hours, thereby remaining at the same spot above the earth as it rotates, making the satellite appear stationary from the ground.

This geosynchronous orbit (GEO) allows users to train their antenna at a particular spot in the sky without any need to track the satellite. Unfortunately, the distance itself is nearly 100 times higher than the orbit of the space shuttle, making launches costly and in-orbit repairs impossible. At this height, light (and radio waves) takes approximately ¼ second to travel from earth to the satellite and back, adding a noticeable delay to voice communications and interfering with data transfer algorithms.

One solution is to move the satellites closer to the ground in LEO (low earth orbit) or MEO (medium earth orbit) orbit. This was the premise behind Iridium and Globalstar, as well as many proposed satellite broadband projects including Teledesic. All of these projects failed.

Because LEO and MEO satellites move in relation to the ground, a fleet of satellites is required to keep at least one satellite in view at all times. The lower the orbit, the more satellites are needed. Iridium, for example, consists of a constellation of 66 satellites. Not only are large numbers of satellites required, but the full constellation must be in orbit before the system can provide continuous service. In contrast, a GEO satellite can provide coverage to users over one-third of the earth with the launch of a single satellite.

Further, due to the movement of LEO satellites relative to users, sophisticated hand-off mechanisms are necessary to transfer users between satellites as they disappear over the horizon. On the ground, unidirectional antennas are sufficient for telephony, but broadband applications require an antenna that can track rapidly moving satellites, making customer premise equipment prohibitively expensive for consumers.

The satellite phone systems do offer basic data connectivity, with data rates limited to 2.4 Kbps for Iridium and 9.6 Kbps for Globalstar. Though hardly a replacement for broadband, with worldwide coverage and not even an antenna to install, satellite phones can be convenient for text messaging and occasional text-based e-mail, especially in emergency situations.

Internet Protocols

TCP/IP (Transmission Control Protocol/Internet Protocol) is the suite of protocols that computers use to communicate with each other over the internet and most other computer networks. Historically, there have been many other protocols, both proprietary and standardized, but the growth of the internet turned TCP/IP into the dominant standard and made computer networking and TCP/IP synonymous with each other.

The Internet Engineering Task Force (IETF) is the organization responsible for the standardization of TCP/IP and associated internet protocols. Unlike most standards bodies, the IETF is open, with no official membership. Anyone interested may participate and even vote. While the organization holds large meetings three times per year, much of its work creating standards takes place on subcommittee mailing lists. Rather than formal voting procedures, the IETF operates on the principle of "rough consensus and working code" to validate that the ideas proposed on paper interoperate in practice. IETF standards are published as numbered Request for Comments (RFC), but it is important to note that most RFCs are informational and do not denote standards.

TCP/IP consists of a large number of component protocols operating at different layers to manage data transfer across computer networks. The most fundamental of the protocols are IP, which handles routing of data across the network, TCP for the reliable transfer of data between end nodes, and UDP for real-time streaming.

IP (Internet Protocol)

IP is responsible for moving packets of data across a computer network from source to destination. IP packets can carry any digitized content; not only web or e-mail but voice, video, fax, music, and even television.

Every machine on the Internet is assigned a unique 32-bit IP address, usually written in decimal form of 192.168.18.5. Data is broken into small packets, usually 1500 bytes each, at the source. Each packet has a header, which can be thought of as the envelope on a letter, listing the address of the destination along with other control information. Routers at each intersection of the network look up in a table that the destination address on each packet.

IP is stateless, examining each packet individually though it might be part of a larger flow of data between two end points, and quite simple. It does no prioritization, treats every packet equally, and knows nothing about the data or applications riding inside the packets. In contrast to traditional, circuit-based telephony which depends on large, centralized switches, IP was designed to keep the network simple and flexible. When a link becomes congested, the router queues a small number of packets, but any excess beyond that is thrown away.

The challenge to IP is building the tables that tell the router which direction to send each packet. For small networks, this table listing the individual machines on the local network can be entered by hand. Any packet with a destination address in the table is forwarded to the local network. Everything else is sent to the next router on the internet. The router doesn't need to know the path the packet takes to reach its final destination. Each router along the route looks up the address in a table and forwards it to the designated next router.

Of course, big routers at ISPs and on the internet backbone handle a huge volume of traffic and have to maintain complex tables that change frequently as new networks are brought on-line. These tables are impossible to create by hand, and IP has mechanisms for routers to exchange information between each other to determine the best paths to use and automatically build the routing table.

IP can run over any type of network, whether terrestrial, wireless, or satellite, high-speed or low-speed. Any type of network can carry IP traffic, one important factor in the immense growth of the internet. With no need to change the protocol or even update the software on the end nodes, IP has been able to take advantage of advances in technology, from high-speed Ethernet to DSL to wireless communications, making the internet extensible and future-proof.

The one significant limitation of the current version of IP, known as Internet Protocol version 4 (IPv4) is the number of unique addresses it can support. Using 32 bit addresses, IPv4 is limited to 4.3 billion addresses, and with parts of the address space reserved, the usable total is considerably less. This was more than sufficient when the internet was a small network for engineers and scientists, but with the internet now a global, all-inclusive network connecting not only computers, but Personal Digital Assistants (PDAs), mobile phones, game consoles, televisions, and eventually even home appliances, IPv4 will run out of addresses. The IETF has therefore developing a new version, designated IPv6, which expands IP addresses to 128 bits.

Unfortunately, IPv4 and IPv6 are not compatible. All computers and applications will therefore have to be updated to IPv6 eventually. Networks will have to be redesigned and renumbered, a tedious, time consuming, and expensive project, with no quantifiable benefits to users. For this reason, deployment of IPv6 has been slow. However, the needs of the U.S. military and the growth of IP-enabled mobile phones is starting to drive deployment, with the expectation that IPv6 will start to see active usage from 2008.

TCP and UDP

While IP handles the movement of individual data packets across the network, two higher-layer protocols, UDP and TCP, are responsible for the end-to-end communications between the source and destination devices.

Data is broadly classified as either real-time or reliable. UDP (User Datagram Protocol) is used for most real-time applications such as voice and video where it is more important for the packets to arrive quickly than to guarantee that all packets arrive. TCP (Transmission Control Protocol) is used for traditional data applications such as e-mail, web, database and file transfer where it is necessary to ensure that all data arrives at the destination intact and in its original order.

UDP is a simple protocol running over IP, with no feedback, reliability or self-throttling mechanisms. Data is streamed across the network and any lost packets are ignored. With no timing mechanisms, UDP is not sensitive to satellite latency. However, for the applications themselves, particularly voice, the quarter second satellite delay can be noticeable to users. Unfortunately, until a way is invented to increase the speed of light, this delay is unavoidable when using GEO satellites. For good performance, it is therefore critical to minimize any other delays in the network.

TCP is a highly complex protocol used for the reliable transfer of data between two end nodes over an unreliable IP network. At the receiver, TCP reassembles the packets to their original order. Acknowledgements allow the sender to keep track of which packets arrived and retransmit any that are lost or corrupted. The flow of acknowledgements from the receiver is also used to throttle the sender's transmission rate when the network is congested.

Unfortunately, TCP assumes that any delay in the return of acknowledgements is caused by network congestion and consequently, satellite latency causes TCP to limit its transmission rate, often leaving most of the bandwidth unused. Similarly, TCP assumes that any packets that are lost are due to network congestion rather than corruption of the data, and cuts its transmission rate sharply, which can have a disastrous effect on performance over high-error satellite links.

These issues have historically limited TCP throughput over satellite, causing slow downloads and wasting expensive bandwidth, branding satellites as slow and inappropriate for internet applications. Fortunately, the problem is not the satellite latency itself but TCP's reaction to the delay, and a variety of acceleration techniques are now commonly employed to overcome these limitations, making satellite performance equivalent to terrestrial networks.

IP Acceleration

To overcome the limitations of TCP, satellite data networks require TCP acceleration functionality called Performance Enhancing Proxy (PEP) or Protocol Gateways. PEPs work by intercepting TCP connections on the local network and transmitting the data over the satellite using a protocol optimized for long delay and high error conditions. On the other side of the link, another PEP converts the traffic back to TCP, rendering the process transparent to end users. PEP functionality can be built into the satellite terminal, added to the network with a separate acceleration device, or loaded onto the PC as client software.

To further enhance performance and conserve bandwidth, most satellite acceleration systems also include compression functionality. While the PEP allows TCP to take full advantage of available bandwidth and is therefore most useful on unsaturated links, compression actually saves bandwidth and provides benefits on links that are already saturated. Lossless data compression reduces the amount of data transferred over the link by finding repeated patterns in the data and replacing them with codes. On the other side of the link, the decompression process reconstructs the original data. Compression generally provides the greatest benefits for text-based materials. Graphics are usually transmitted in pre-compressed formats such as GIF or JPEG and do not see any benefit from compression.

A similar technique, header compression, replaces the TCP, UDP and IP headers with a short code. For full-sized, 1500 byte packets, header compression only saves a few percent of the bandwidth, but real-time applications, especially Voice over IP (VoIP), transmit large numbers of very small packets. With a 48 byte payload, reducing 40 bytes of header down to only 4 bytes increases the number of calls that can be carried over a fixed amount of bandwidth by 70%. However, once the header has been compressed, it can not pass through another router. Header compression can therefore only be performed on the last networking device prior to the satellite link and must be decompressed by the first networking device on the other side.

Satellite internet systems, especially those designed for consumers, usually also include HTTP prefetching to improve web browsing speeds. HTTP prefetching proactively retrieves the individual objects on each web page and pushes them across the satellite link, making them available locally when the browser requests them. HTTP acceleration is especially effective for improving the responsiveness of static web pages with large numbers of embedded graphic elements, and is often combined with caching to conserve bandwidth.

Additionally, some applications such as Windows file sharing have their own delay-induced performance limitations, and a variety of vendors offer proxies to improve performance of each application.

Satellites and the Internet

Broadband Satellites

New broadband satellites utilizing spotbeam technology promise much lower bandwidth costs, thereby overcoming the primary handicap of satellites towards broader utilization. Traditional satellite transponders cover an entire continent, ideal for broadcasting a television signal over a wide area, but wasteful and expensive for transmitting data or phone calls to individuals. In contrast, spotbeams, usually in the higher frequency Ka-band, focus the satellite transponder on a local region the size of a city. The same frequency can be re-used across multiple transponders, thereby multiplying the total bandwidth capacity of the satellite and sharply reducing bandwidth costs.

The two most prominent next-generation satellite systems are WildBlue and IPStar.

WildBlue

WildBlue finally launched its long-anticipated satellite broadband service in June 2005, offering high-speed internet service comparable to DSL and cable modems to homes and small offices in the continental United States. Packages prices are US $49.95/month for 512 Kbps and US $79.95/month for 1.5 Mbps, with installation and equipment costing US $479.

The WildBlue service uses the Ka-band transponders aboard Telesat's Anik F2, though WildBlue is also constructing their own dedicated satellite for later launch. The network is based on the DOCSIS standard, with equipment manufactured by ViaSat. Investors in WildBlue include many of the biggest names in satellite and media communications: Intelsat, Telesat, National Rural Telecommunications Cooperative (NRTC), and Liberty Media.

At least initially, WildBlue is focusing its marketing efforts on small towns and rural areas without access to terrestrial broadband. However, as DSL and cable continue to expand their rollout and new WiMax equipment brings simple and inexpensive wireless coverage to entire communities, WildBlue will need to compete with terrestrial offerings based on price and service. WildBlue is therefore seen as a bellwether for the satellite industry, and if successful, is likely lead to the development of similar projects in other parts of the world.

171

The Satellite Technology Guide for the 21st Century

IPStar

In the Asia-Pacific region, Shin Satellite, a Thailand-based satellite operator, is developing the iPSTAR satellite broadband system. The IPStar satellite, launched in 2005, has 84 Ku-band spotbeams with a total capacity of 45 Gbps covering 22 countries throughout the region, as much bandwidth as all other satellites in Asia combined.

The IPStar service is aimed at a wide variety of users, from individual consumers to large corporations and government organizations. Service offerings range from consumer broadband to corporate virtual private networks, video on demand, and telephony.

The system uses Ku-band spotbeams for communications to user terminals, avoiding the rain fade limitations of Ka-Band in a region prone to frequent heavy storms, while higher-powered feeder beams between teleport and satellite take advantage of the bandwidth capacity of Ka-band. The forward channel of up to 45 Mbps is based on a proprietary TDM-OFDM technology while the return channel uses MF-TDMA for bursty traffic and dedicated bandwidth for high data rate applications.

Prior to launch of the satellite, the IPStar service was rolled out across Asia-Pacific using Ku-band transponders aboard existing Thaicom satellites.

Standards Wars: DVB-RCS vs. DOCSIS

Standardization drives price competition, and lower prices open the technology to new markets, thereby increasing volumes, driving economies of scale and dropping prices further in a virtuous circle. The satellite industry, long dominated by expensive, proprietary equipment, is looking towards standardization to help it break free of its traditional niche markets and into broader usage. Two emerging standards, DVB-RCS and DOCSIS are vying for supremacy.

DVB-RCS (Digital Video Broadcasting – Return Channel via Satellite) is an open standard designed to be applicable to a wide range of applications, combining a DVB forward channel with an ATM-like satellite return channel. This standard is being developed by SatLabs, a European trade group of satellite operators and equipment manufacturers, which is also responsible for testing to ensure interoperability between equipment from different manufacturers.

The primary benefit to the DVB-RCS standard over traditional proprietary equipment is the interoperability between different head-end and customer premises equipment. In addition to providing flexibility to mix-and-match the best components from different manufacturers, interoperability avoids lock-in to one supplier and the resulting competition is helping to reduce equipment prices.

However, the DVB-RCS standard was developed primarily by equipment manufacturers, each with their own solutions focusing on the needs of different types of customer. The resulting specification is quite complex with many options, not all of which are supported by all vendors. With no large-scale deployment driving production volume, prices for the equipment remain too high for consumers and the system to date has been targeted primarily at small to medium sized businesses.

DOCSIS (Data Over Cable Service Interface Specification), the standard for terrestrial cable modem networks, has been adapted by WildBlue and ViaSat to support satellite communications. The primary benefit of DOCSIS is the ease with which low-cost cable modems can be modified and manufactured, and the availability of off-the-shelf provisioning, monitoring, and billing equipment for operating a large-scale consumer cable modem network.

However, at the current time, DOCSIS modems and satellite head-end equipment are produced only by ViaSat, and there is no independent standards body to ensure interoperability, making it difficult to consider satellite-DOCSIS an open standard.

Nevertheless, DOCSIS equipment seems well-suited to large scale consumer networks, while DVB-RCS fits a need in the higher-speed business market and it seems likely that both will find a niche within their respective markets. It has to be kept in mind, though, that success for either standards effort is not guaranteed, and lower cost, interoperable products by themselves do not necessarily lead to high volumes, especially when the cost of the satellite bandwidth remains the primary impediment towards price parity with terrestrial options. Time will tell whether the combination of equipment standardization and a new generation of satellites can deliver the solution that makes satellites a competitive option for broadband, or whether satellite broadband will remain a niche solution for locations where terrestrial options are unavailable.

CHAPTER 9

The Future of Satellite Communications

Arthur C. Clarke, the discoverer of the geostationary orbit that made artificial satellites possible, formulated the following three laws of prediction:

1. When a distinguished but elderly scientist states that something is possible, he is almost certainly right. When he states that something is impossible, he is very probably wrong.

2. The only way of discovering the limits of the possible is to venture a little way past them into the impossible.

3. Any sufficiently advanced technology is indistinguishable from magic.

As we have seen, satellite communications technology has developed more rapidly and far greater in scope that Clarke imagined when he initially conceived the idea in 1945. The commercial satellite industry, which only started in 1964, has grown to the US$ 100 Billion-a-year, global industry that is today.

So, what future does the rest of the 21st century hold for the satellite communications industry? I hesitate to make any firm prognostications, for as one wag put it, it's easy to predict what the future holds—the only question is *when*?

However, if past performance is any indicator, the future bodes well for the satellite industry. In the last ten years when the Satellite Industry Association (SIA) and research firm Futron began tracking the global satellite business, the industry has posted consistent average growth rates of 10 percent per annum even during the years considered to have been one of the worst downturns in the telecommunications industry. This rate of growth is projected by analysts to continue in the coming years.

The leading research companies monitoring the satellite industry, notably NSR(www.nsr.com). Euroconsult (www.euroconsult-ec.com),Futron Corporation (www.futron.com) and the Teal Group (www.tealgroup.com), are all very bullish in their recent reports and market studies about the medium- and long-term prospects of the industry.

One key endorsement of the satellite industry has been the influx of private equity firms (PEFs) that have been buying up satellite companies since the highly publicized sale of PanAmSat in 2004 by the leverage firm of Kohlberg Kravis Roberts & Co. That led to many other acquisitions of satellite operators by PEFs, who are now moving in on satellite ground equipment companies and service providers. The consensus among industry analysts on the impact of PEFs on the industry, is that it has been largely positive and that it serves to affirm the long-term financial viability of the industry. PEFs have also instilled more fiscal discipline to the companies it has taken over, which has affected other companies as well.

The Teal Group believes the commercial satellite industry is on the verge of a growth cycle that might extend until well into the next decade. Teal estimates orders for more than 200 geostationary commercial communications satellites worth more than $25 billion through 2016. Orders for more than 100 low earth orbit mobile communications replacement satellites are expected to generate up to $4 billion in business. Next-generation U.S. military satellite programs are estimated at a total cost of $100 billion.

There is much cause for optimism. The world has changed significantly since the first satellite organization Intelsat was formed a little over forty years ago, with a lot of the changes directly attributable to the impact of satellite technology. Consumers today interact with various media and demand information very differently. As former SIA executive David Cavossa put it, the market today constantly demands "more power, more mobility and more convergence." The consumer mantra today seems to be to have access to information "anytime, anywhere and through any device." The increasing demand for bandwidth-hungry applications such as IPTV, Mobile TV and HDTV will require network solutions that will have to consider satellite technology as part of the equation to meet growing requirements.

One of the most dramatic growth came from HDTV channels which grew from 67 at the end of 2005 to 471 in May 2007—a 600 % increase in less than 18 months. Even with such an increase, HDTV channels still account for only 2% of all television channels carried by satellites today—so this growth trend is expected to continue.

Demand for commercial satellite services will also come from the government and military sectors, which are expected to continue their reliance on the commercial sector for its growing needs. Research and Markets (www.researchandmarkets.com), which tracks US Defense spending, sees substantial increases every year and forecasts US defense spending on space to reach US $28 Billion in 2010, from US$ 15 Billion in 2000. More than $ 18 Billion is spent annually on the development of space systems. Research firm NSR forecasts that demand for commercial satellite capacity by the US government will enjoy double digit growth through 2012. The US Defense Information Systems Agency (DISA), which is responsible for purchasing commercial satellite capacity for the Department of Defense (DoD), has been spending between $300-350 million annually since 2003, up from less than $100 million in 2000. The US DoD is the largest government procurer of commercial satellite services in the world today.

Demand for satellite services might potentially come from other sectors. Aviation experts say that the US air traffic system developed in the 50s will require a US $40 Billion upgrade—a substantial portion of which will involve the use of satellite-based technology. Growing concerns over cross-border security would require upgrades of national databases and networks to facilitate the exchange of information between countries. The Global Positioning System (GPS), developed by the US in the 70s, is also in need of an upgrade.

One market segment that has not been fully tapped by the satellite services sector is the enterprise market. With an increasingly global economy where literally millions of companies need to link up with their many branches and a growing number of telecommuting workers worldwide, the prospects for growth in this market are very promising.

The Satellite Technology Guide for the 21st Century

One other sector that is not in everyone's radar is space tourism. After the successful flight in 2004 from the Mojave desert in California of the first privately-funded manned spacecraft, SpaceShipOne, several commercial ventures have been launched with the aim of making space travel available to the general public. Most notable of these ventures is Virgin Galactic, backed by billionaire mogul Sir Richard Branson. Virgin Galactic announced their intention to have inaugural flights to space in 2009 at a cost of between US$ 100-200K per person.

The successful flight of SpaceShipOne, earned its designer Burt Rutan the US$ 10 million Ansari X-Prize, patterned after the Orteig Prize won by Charles Lindbergh in 1927 for his first solo flight across the Atlantic. If space travel ever takes off in the 21st century at even a fraction of the rate that air travel grew in the early 20th century, there will certainly be a spillover effect on the satellite industry. Space tourism will require a lot of the same technology that the satellite industry has been developing and providing through its manufacturing, equipment and services sectors.

There certainly is no dearth of emerging applications, services and markets that will provide opportunities for the satellite industry in the years to come. These opportunities, coupled with the continuing innovation and development of satellite technologies, will ensure the long-term viability of the commercial satellite industry well into this century.

It is very likely that we will continue to see the development of the satellite spacecraft with more on-board processing capabilities, higher power and larger aperture antennas that will enable satellites to handle more bandwidth, promote frequency re-use and act like "super computers in the sky." Future satellites will be able to refocus spot beams not just to respond to market demands, but to requirements at different times of the day. We will also see further improvements in a satellite's propulsion and power systems which will enable satellites to increase their service life to 20-30 years from the current 10-15 years.

In addition, to more powerful and capable satellites, other technical innovations are in the works include the following:

• NASA, in cooperation with private companies such as Dulles, VA-based Orbital Sciences Corporation, is developing low-cost, reusable launch vehicles which will be the prototype for future launchers. NASA

The Future of Satellite Communications

pioneered a reusable launch vehicle in the 80s with the Space Shuttle, but it has not been a cost-effective alternative to conventional satellite launches. The development of low-cost reusable launchers will mean substantial savings in the cost of satellite launches—one of the most expensive items in a satellite venture.

- The U.S.' Defense Advance Research Projects Agency (DARPA) is testing the feasibility of breaking down the key parts of a satellite and launching each of those components separately into orbit. A demonstration mission is planned for 2010 or 2011. If successful, launching a satellite's components separately would prevent the catastrophic loss of an entire US$ 200-million satellite in case of launch failure. Launching a satellite in separate modules will spread the risk of failure to several launches instead of one, making a loss of one component not as devastating as a total loss of the entire satellite, as in the case of launches today. This would mean that operators would not have to rebuild another satellite which could take up to two years, and just replace the component that was lost, saving valuable time, not to mention cost.

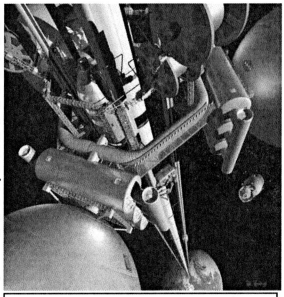

An artist's rendering of the proposed space elevator. (Image courtesy of NASA)

- An even more revolutionary idea for launching payloads into space is the space elevator, which would consist of a cable anchored to the earth's surface, reaching into space. The space elevator would preclude the need for launch vehicles altogether. It has many adherents, including Arthur C. Clarke, who originally came up with the idea in 1979 in his science fiction work, *The Fountains of Paradise*. The idea is actually based on sound physics and has been endorsed in several scientific conferences and supported financially by grants from NASA.

The Satellite Technology Guide for the 21st Century

The space elevator requires a super-strong cable made from high-tech materials extending from the earth's surface to beyond the geostationary arc of 22,300 miles. By affixing a counterweight at the end of the cable or by further extending the cable, inertia ensures that the cable remains stretched and taut, countering the gravitational pull on the lower sections. This ensures that the elevator will remain in geostationary orbit. The installation of the space elevator will be like a direct railway line to space and would greatly diminish the cost and increase the reliability and frequency of space launches. It will eliminate the need for massive rockets used to escape the earth's gravitational pull and it will be able deliver payloads to geostationary orbit with lesser risk. The idea might seem like stuff of science fiction, but coming from Clarke, who came up with the idea of communications satellites himself, it might not be a good idea to bet against it.

Others means of launching objects into space without using large amounts of fuel unlike current rocket-based launchers, are being explored. One concept propounded by Dr. Young K. Bae in Southern California involves the use of laser technology to launch objects into space dubbed "photonic laser thrusters." His research has been partially funded by NASA and he is talking with private companies to develop the technology.

As Clarke so aptly put it in his second law of prediction, "the only way of discovering the limits of the possible is to venture a little way past them into the impossible." Looking back at the history of the development of satellite technology, it would appear that the industry was guided by this very principle. To have achieved what it had in such a relatively short span of time, demonstrates a capacity for taking big risks and not be hindered by the limitations of their imaginations, energies and resources. Indeed, the satellite industry is living proof that not even the sky is the limit to what it can accomplish.

APPENDIX 1
GLOSSARY OF TERMS AND ACRONYMS

A

Amplifier-An input/output device which increases the voltage, current or power level of a signal.

Analog-A type of signal transmission that is susceptible to changes in light, sound, heat and pressure.

ADC (Analog-to-Digital Conversion)-The process of converting analog signals to digital signals. The reverse process is called DAC or Digital-to-Analog Conversion.

AKM (Apogee Kick Motor)-Rocket motor fired at the highest point of a satellite's orbit to put it in its proper geostationary orbit.

Antenna-A device for transmitting and receiving radio waves. In satellite communications systems, an antenna is part of an earth station's sub-system.

Antenna Farm-The area of an earth station or teleport where all the antennas are located.

Aperture-The area of an antenna that can receive a satellite signal.

Apogee-The point in an elliptical satellite orbit which is furthest from the surface of the earth.

Atmospheric Noise-Radio noise caused by natural phenomena such as lightning and thunderstorms.

Attenuation-The resulting loss in power of electromagnetic signals from transmission to reception.

Attitude Control-The control of the orientation of a spacecraft relative to a gravitating body influencing its flight path. With geostationary satellites, attitude control refers to the orientation of the satellite relative to the sun and the earth.

The Satellite Technology Guide for the 21st Century

ATSC (Advanced Television Systems Committee)-Industry group, established in 1982, that develops standards for digital and high definition television.

B

Backhaul-A terrestrial communications channel linking an earth station to a local switching network.

Bandwidth-A measure of spectrum or frequency use or capacity. Video and audio signals are measured in Hertz (Hz) while data is measured in Bytes.

Baseband-The frequency band which contains the basic low frequency information from a primary source like a television camera, satellite receiver, or recording device before modulation or demodulation.

Bird-Colloquial term for a communications satellite.

Bit-A single digital unit of information.

BPS (Bits per second)-A measure of the speed of data transmission.

BPSK (Binary Phase Key Shifting)--a digital modulation technique to compress signals.

Bit Rate-The speed of a digital transmission, measured in bits per second.

Body-stabilized satellite-also known as the "three-axis" stabilized satellite where the satellite's position is maintained in orbit by making adjustments on three axis of the satellite as opposed to a spin-stabilized satellite that maintains the position of the satellite by spinning the body like a top.

BOL (Beginning of Life)- The beginning of a satellite's projected service life in orbit, which currently averages 10-15 years for geostationary satellites. A satellite at its BOL is more reliable and can accommodate longer-term leases than a satellite at its EOL (End of Life).

Broadband-High speed digital transmission of voice, data and video signals.

Glossary

BSS (Broadcast Satellite Services)-ITU designation for video and audio services on satellites.

Bus-The part of the satellite that houses the power, propulsion, thermal, telemetry and control systems.

C

Carrier-A radio frequency signal that is modulated with an input signal for the purpose of conveying video, audio and data signals. A transmission can have a main carrier which is the video and several sub-carriers which could contain the audio and data.

C-Band-The frequency band between 4 and 8 GHz most commonly used in satellite communications.

C/N (Carrier to Noise Ratio)-The ratio of the signal carrier to the noise carrier in a given signal. Generally, the higher the C/N, the better the quality of the reception.

CCITT (Comité Consultatif Internationale de Télégraphique et Téléphonique)-An International body, affiliated with the ITU, which establishes worldwide standards for telecommunications. Reorganized to include CCIR (Comité Consultatif Internationale de Radio) and renamed ITU-T (Telecommunications Standardization Sector).

CDMA (Code Division Multiple Access)-A multiple-access system which uses spread-spectrum modulation.

Channel-A frequency band in which a specific signal is transmitted.

Circular Polarization-A rotating pattern of transmission that enables two different signals to be transmitted to the same frequency. A circularly-polarized transponder can receive both right-hand rotating and left-hand rotating signals on the same frequency, doubling its capacity.

Clarke's Orbit-The circular orbit in space 22,300 miles from the surface of the earth where geostationary satellites are located. Named by the International Astronomical Union after renowned science fiction writer Arthur C. Clarke, who came up with the idea of geostationary satellites in 1945.

The Satellite Technology Guide for the 21st Century

Co-Location-The sharing by multiple satellites of the same geostationary orbital assignment made possible by differentiating the frequency bands in which each satellite operates.

Compression-The digital process of reducing the quantity of data used to represent video, audio, voice and other data, with the goal of retaining as much of the original's quality as possible. The main purpose of compressing data is to save on bandwidth, since a compressed signal requires less bandwidth to transmit.

Conditional Access-A hardware or software system managing access to signals.

CONUS (Contiguous United States)-All the states in the continental U.S. which excludes Hawaii and Alaska.

Coverage Area-The service area wherein a satellite's signal can be received on the ground. Also known as a satellite's "footprint."

Cross Strapping-The ability of a satellite's transponder to receive a signal in one frequency band e.g. C-Band, and to retransmit that signal to another frequency band e.g. Ku-Band.

D

DAC (Digital-to-Analog Conversion)-The process of converting digital signals to analog signals. The reverse process is called ADC or Analog-to-Digital Conversion.

DAMA (Demand Assigned Multiple Access)- A method of sharing bandwidth in a transponder based on user demand.

DBS (Direct Broadcast Satellite)-Satellites used exclusively to broadcast multiple channels of programming directly to consumers' homes. Examples of DBS satellites are the DirecTV and Echostar fleet of satellites in the US.

dBW-The ratio of the power of a signal to one Watt expressed in decibels.

Decibel (dB)-A unit of measuring a signal's power named after telephone pioneer Alexander Graham Bell. A decibel is one-tenth of a Bel.

Glossary

Demodulation-A process where the original modulated signal is recovered from a modulated carrier.

Digital-A type of signal transmission that converts information i.e. video, audio and data into bits of 1s and 0s. Digital transmission lends itself better to compression of the signal to save on bandwidth and is not as susceptible to noise and interference as analog signals.

Dish-Another term used to refer to an antenna which is capable of receiving and/or transmitting satellite signals.

DMB (Digital Multimedia Broadcasting)-A digital radio transmission standard for sending multimedia (radio, TV, and datacasting) to mobile devices such as mobile phones.

Downlink-The transmission path from the satellite to the ground facility or earth station.

DTH (Direct-to-Home)-Satellite service that provides multichannel offerings direct to the consumer's home.

DOCSIS (Data Over Cable Service Interface Specifications)-An international standard developed by CableLabs for a data interface over cable systems.

Dual-Band-In antenna systems, dual-band antennas have the ability to receive or transmit in two different frequency bands e.g. Ku-Band and/or C-Band.

DVB (Digital Video Broadcasting)- The European initiative to promote adoption of digital video standards. Some of the standards the DVB has promulgated include: DVB-RCS (Return Channel Satellite-used for VSAT and data networks); DVB-T (Terrestrial); DVB-H (Handheld), DVB-S (Satellite); DVB-SH (Satellite Handheld) and others.

E

Earth Station-Equipment consisting of an antenna, low-noise amplifier (LNA), downconverter, and other ancillary equipment capable of receiving and/or transmitting satellite signals.

The Satellite Technology Guide for the 21st Century

Edge of Coverage-Limit of a satellite's defined service area.

EIRP (Effective Isotropic Radiated Power)-The power of the signal from the satellite antenna or the transmitting earth station antenna. The transmit power in units of dBW is expressed by the product of the transponder output power and the gain of the satellite transmit antenna.

Electromagnetic Spectrum-the range of all possible electromagnetic radiation occurring in nature which includes radio waves.

Encoder -A device used to electronically alter a signal so that it can only be viewed using a receiver equipped with a special decoder.

Encryption-The process of scrambling a signal so that only authorized users with the proper decoding equipment can unscramble the signal.

EOL(End of Life)- The last years of a satellite's projected service or operational life of between 10-15 years.

F

Fairing-The part of a launcher that contains the payload or the cargo that the launcher has to launch into space.

FCC (Federal Communications Commission)-The U.S. federal body responsible for regulating communications, incluuding satellite communications.

FDMA (Frequency Division Multiple Access)- The use of multiple carriers within the same transponder where each signal has been assigned a frequency slot and bandwidth.

Feed- A term that refers to a signal source, for example, the Discovery video channel may be referred to as the "Discovery Channel feed."

Flyaway-A mobile Satellite News Gathering (SNG) or satellite communications equipment that can be transported from one location to another.

Footprint- The coverage area of a satellite.

Forward Error Correction (FEC)-A system of error control for transmissions of compressed and encrypted digital signals.

Glossary

Frequency-The number of wavelengths or cycles per second of a particular radiation or radio wave measured in Hertz (Hz). One cycle per second is 1 Hz, 1000 cycles per second, one kilohertz (KHz); 1,000,000 cycles per second, one megahertz (MHz): and 1,000,000,000 cycles per second, one gigahertz (GHz). Satellite transmissions occur in the 1-50 GHz frequency range.

Frequency Reuse-The use of different polarities (such as vertical and horizontal polarization) transmitting in the same frequency in order to maximize the bandwidth of a communications satellite.

FSS (Fixed Satellite Services)-A category of satellite broadcasting and telecommunications services to distinguish them from DBS (Direct Broadcast Services) and MSS (Mobile Satellite Services).

G

Gain -A measure of amplification expressed in dB.

Geostationary orbit-The orbit approximately 22,237 miles directly above the earth's equator (0° latitude) discovered by Arthur C. Clarke in 1945. In this orbit, a satellite will move as the same speed as the earth's rotation, which makes the satellite's position fixed relative to a position in the ground.

Geostationary Transfer Orbit-an elliptical orbit where a satellite is placed during the launch process just before it is propelled into its proper geostationary orbit.

Geosynchronous orbit- Same as geostationary orbit, with the exception that certain inclined orbits may qualify as geosynchronous orbits but not as geostationary orbits, due to the slight deviations in its orbital position.

GPS (Global Positioning System)-A system of satellites dedicated to providing precise positioning on the ground in terms of latitude and longitude. The system was developed by the US military in the 70s.

Ground Segment-The terrestrial or ground component of a satellite network.

G/T (Gain/ Temperature)-A parameter expressed in dB where "G" is the net gain of the signal and "T" is the noise temperature of the signal.

H

HDTV (High Definition Television)-A digital broadcasting standard for high quality TV reception.

Hertz (Hz)-A unit of measurement of frequency.

HPA (High Power Amplifier)-Equipment that boosts signals to very high RF power. In satellite communications, there are three types of HPAs used: Solid State Power Amplifiers (SSPA), Klystron Power Amplifiers (KPA), and Travelling Wave Tube Amplifiers (TWTA).

Horizontal Polarization-Transmission of linearly polarized radio waves whose electric field is parallel to the earth's surface.

Hops-The number of receive and transmit points during transmission from one location to another.

Hybrid Networks-Communications networks using a variety of media which can include a mix of satellite, terrestrial and wireless technologies.

I

IF (Intermediate Frequency)- A frequency used within ground equipment as an intermediate step in transmitting or receiving a signal. In earth stations and teleports, the most commonly used IF carrier frequency is 70 MHz.

Inclination-The angle between the orbital plane of a satellite and the equatorial plane of the earth.

Inclined Orbit-A condition that occurs when a satellite's position is no longer corrected along the north-south direction. A satellite operator might do so to extend the life of a satellite by saving on fuel used to make periodic adjustments in the satellite's position. The inclination happens gradually over time.

In-orbit Testing-The period after a satellite is successfully placed in its correct orbital position where it undergoes extensive tests before it starts operation.

Interference-Unwanted electrical signals or noise causing degradation of reception.

Glossary

Inter Satellite Links (ISL)- Radio or optical communications links between satellites.

IRD (Integrated Receiver Decoder)-A receiver of satellite signals that decodes encrypted or decompresses compressed signals.

IP (Internet Protocol)-Standards by which data is sent from one computer to another on the Internet.

IPTV-A two-way digital broadcast signal delivered over IP networks.

ISP (Internet Service Providers)-Companies providing network access to the Internet for business, government and consumer markets.

ITU (International Telecommunication Union)-The international regulatory body for telecommunications, which includes satellite communications, based in Geneva, Switzerland.

K

Ka-Band-The frequency range from 18 to 31 GHz.

Klystron-A microwave tube which uses the interaction between an electron beam and the RF energy on microwave cavities to provide signal amplification.

Ku-Band -The frequency range from 10.9 to 17 GHz.

L

Latency-Refers to the characteristic of satellite signals which produces delays in transmission and reception due to the vast distances the signal has to travel.

L-Band -The frequency range from 0.5 to 1.5 GHz.

Link Budget-The calculation of power and noise levels between transmitter and receiver (uplink or downlink). Factors considered include antenna size, satellite transmission power and potential atmospheric effects (from weather to sunspots).

Look Angle-The elevation and azimuth at which a particular satellite is predicted to be found at a specified time. This parameter is used to

The Satellite Technology Guide for the 21st Century

determine the exact pointing of an antenna to a satellite, since an antenna has to be directly looking at a particular satellite to be able to receive or transmit signals to that satellite.

LNA (Low Noise Amplifier)-The preamplifier between the antenna and the earth station receiver. The LNA amplifies the weak signals received from the satellite.

LNB (Low Noise Block Downconverter)-A combination Low Noise Amplifier and downconverter built into one device attached to the receive port of an antenna.

LEO (Low Earth Orbit)-Orbits at an altitude of 100-200 miles from the earth.

M

Master Control-The nerve center of an earth station or teleport where all the signals being transmitted and received are monitored and key equipment are controlled.

MCPC (Multiple Carrier Per Carrier)-A signal comprised of multiple digital streams that are multiplexed into a single stream, which is then transmitted on a single carrier. MCPCs are typically used to combine multiple compressed digital video signals into one carrier or transponder to maximize bandwidth use.

MEO (Medium Earth Orbit)-Orbits at an altitude of 1000 miles from the earth.

Microsatellites-Satellites weighing less than 100 kg. (220 lbs).

Modulation-The process of manipulating the frequency or amplitude of a carrier in relation to an incoming video, voice or data signal.

MPEG (Motion Pictures Expert Group)-A committee first formed in 1988 to develop international standards for video and digital storage. MPEG standards cover a range of compression formats such as MPEG-2 and MPEG-4.

Multiplexing-Techniques that allow a number of simultaneous transmissions over a single circuit.

Glossary

Mux-Short for Multiplexer. Combines several different signals (e.g. video, audio and data) into a single communication channel for transmission. Demultiplexing separates each signal at the receiving end.

MSS (Mobile Satellite Services)- Services transmitted via satellites to provide mobile telephone, paging, messaging, facsimile, data, video and other services directly to users with mobile devices.

N

Nanosatellites-Satellites weighing less than 10 kg. (22 lbs).

NASA (National Aeronautics and Space Administration)- The U.S. agency which administers the American space program.

Network- A communications system linking several points or locations.

Noise-Any outside signal or background atmospherics that interferes with a signal.

NTSC (National Television Standards Committee)-A television standard established by the United States and adopted by numerous other countries. Europe developed a competing standard called "PAL" (Phase Alternation System).

O

On Board Processing-The ability of a satellite to perform many operations including switching and rerouting of signals and error correction, among others.

Orbit-The path that a satellite makes around the earth under the influence of a force such as gravity.

Orbital Position-Specific location of a satellite in the geostationary arc, specified in degrees, east or west longitude.

Operational Life-The service life of a satellite, typically 10-15 years for geostationary satellites.

P

PAL (Phase Alternation System)-A television standard developed in Europe that is incompatible with the US' NTSC television standard.

The Satellite Technology Guide for the 21st Century

Payload-In the satellite spacecraft, the payload is the part that houses the communications systems, including the transponders and the antennas. In launch services terminology, the "payload" is the cargo that launchers carry into space.

Perigee-The point in an elliptical satellite orbit which is closest to the surface of the earth.

Perigee Kick Motor (PKM)-Rocket motor fired during the launch process to place a satellite into a geostationary transfer orbit from a low earth orbit.

Picosatellites-Satellites weighing less than 1 kg. (2.2. lbs.).

Point-to-Point Communications-Communication between one source to a single receiver, for example, a telephone call.

Point-to-Multipoint Communications-Communication between one source to many receivers such as a television broadcast. It is widely accepted that satellite technology is best suited for point-to-multipoint communications because of its wide coverage area.

Polarization-The property of electromagnetic waves, such as radio waves, that describes the direction of the electric field. Two signals can be transmitted in the same frequency without interfering with each other by varying the polarity of the signals. Satellite transmissions can be transmitted with horizontal and vertical polarization as well as right-hand or left-hand polarization. Polarization increases the bandwidth capacity of satellite transponders by reusing frequencies.

Pre-emptible transponder-A type of transponder leasing arrangement wherein an operator can reclaim the transponder in cases of emergencies or other conditions.

PTT (Post, Telephone and Telegraph)-The national regulatory bodies governing telecommunications.

Q

QAM (Quadrature Amplitude Modulation)-A method of combining two Amplitude Modulated (AM) signals - each having the same frequency,

Glossary

but differing in phase by 90 degrees, into a single channel, thereby doubling the effective bandwidth.

QPSK (Quadrature Phase Shift Keying)-A digital modulation technique in which the carrier phase takes on one of four possible values.

R

Rain Fade-Loss or degradation of signals at Ku- or Ka-Band frequencies caused by heavy rainfall.

RF (Radio Frequency)-A range of frequencies in the electromagnetic spectrum that is higher than the audio frequency but below the infrared frequencies. Satellite communications fall under the radio frequency spectrum.

Router-Network layer device that determines the optimal path along which network traffic should be forwarded. Routers forward packets from one network to another based on network layer information.

S

SCPC (Single Channel Per Carrier) A method of transmission in which only one signal or channel is uplinked to a carrier or a transponder.

Set-top Box-An electronic device used to receive digital television transmissions. Set-top boxes can also provide electronic program guides and other data services.

SFD (Saturation Flux Density)-The power required to achieve saturation of a single repeater channel on the satellite.

Signal to Noise Ratio (S/N) -The ratio of the signal power and noise power.

Simulsat-A type of antenna that can receive signals from multiple satellites.

SNG (Satellite News Gathering)-The use of mobile satellite equipment such as a transportable uplink truck to cover live news and special events.

The Satellite Technology Guide for the 21st Century

Solar Outage-Signal outages occurring when an antenna is looking at a satellite, and the sun passes behind or near the satellite and within the field of view of the antenna.

Solar Panels- A device on satellites that converts solar energy into electrical energy using solar cells.

Space Segment-The part of the satellite network based in space consisting mainly of the satellite transponder.

Spectrum-The range of electromagnetic radio frequencies used in video, audio, voice and data transmissions.

Spin Stabilization-A type of satellite stabilization and attitude control which is achieved through spinning the exterior of the spacecraft on its own axis at a fixed rate.

Spot Beam-A signal from a satellite concentrated on a specific geographic area.

SSPA (Solid State Power Amplifier)-Solid state high power amplifier that is gradually replacing Traveling Wave Tube Amplifiers in satellite communications systems.

Stationkeeping-Minor orbital adjustments conducted periodically to maintain the satellite's position.

Subcarrier-A second signal carried onto a main signal. Some satellite transponders carry as many as four special audio or data subcarriers whose signals may or may not be related to the main video carrier.

T

Teleport-A ground-based facility that provides a variety of satellite transmission, program origination, network interconnection and other services.

TCP/IP (Transmission Control Protocol/Internet Protocol)-The suite of communications protocols used to connect hosts on the Internet.

TDMA (Time Division Multiple Access)-A form of a multiple access scheme where a single carrier is time-shared by many users.

Glossary

Three-Axis Stabilization-Type of spacecraft stabilization in which the body maintains a fixed attitude relative to the orbital track and the earth's surface.

Thruster-A small axial jet in the satellite used during routine stationkeeping activities.

Transponder-Communication equipment in a satellite that is a combination receiver, frequency converter, and transmitter.

Turnaround service- Downlinking a signal from one satellite, and retransmitting it to a transponder on another satellite. A service usually provided by teleports.

TT&C (Telemetry, Tracking and Control)-The satellite's control equipment used to monitor on-board satellite operations and to control the satellite's electronics and propulsion equipment.

TVRO (Television Receive Only)- Terminals that use antenna reflectors and associated electronic equipment to receive and process television and audio signals via satellite.

TWTA (Traveling Wave Tube Amplifier)-A type of high power amplifier in a satellite used to amplify the signal before it is retransmitted back to the Earth.

U

UHF (Ultra High Frequency)-The band of frequencies ranging from 300 to 3000 MHz.

Uplink-The transmission path from the ground facility or earth station to the satellite.

V

V-Band-The frequencies in the 37.5-50.2 GHz range. Not currently in use as it is reserved for future demand.

Vertical Polarization-Transmission of linearly polarized radio waves where the direction of the electric field is perpendicular to the earth's surface.

Very High Frequency (VHF)-The range of frequencies extending from 30 to 300 MHz.

VSAT (Very Small Aperture Terminal)-Compact satellite reception and transmission equipment, usually with antennas in the 1.2 to 2.4 meter range.

W

WARC (World Administrative Radio Conference)-A biannual international conference held by the ITU to discuss major issues affecting radio frequency allocation and use.

X

X-Band-The frequency band in the 7-8 GHz range used primarily for military satellite communications.

APPENDIX 2
RECOMMENDED FURTHER READING

Satellite Technology

Elbert, Bruce R. *Introduction to Satellite Communications* 2nd Ed. Boston: Artech House, 1999.

____. *Satellite Communications Applications Handbook.* 2nd Ed. Boston: Artech House, 1999.

Maral, Gerard and Bousquet, Michel. *Satellite Communications Systems.* 4th Edition. New York: John Wiley and Sons, 2004.

Palter, DC. *Satellites and the Internet: Challenges and Solutions.* Sonoma, Calif.: Satnews Publishers, 2003.

Pritchard, Wilbur, Suyderhoud, Henri and Nelson, Robert A. *Satellite Communication Systems Engineering.* Englewood Cliffs, NJ: Prentice Hall, 1993.

History of the Satellite Industry

Bille, Matt and Lishock, Erika. *The First Space Race: Launching the World's First Satellites.* College Station: Texas A&M University Press, 2004.

Gavaghan, Helen. *Something New Under the Sun: Satellites and the Beginning of the Space Age.* New York: Springer-Verlag, 1998.

Labrador, Virgil S. and Galace, Peter I. *Heavens Fill with Commerce: A Brief History of the Communications Satellite Industry.* Sonoma: Satnews Publishers, 2005.

Whalen, David J. *The Origins of Satellite Communications 1945-1965.* Washington, D.C.: Smithsonian Institution Press, 2002.

APPENDIX 3

INDUSTRY RESOURCES

Trade Associations

Asia-Pacific Satellite Communciations Council (APSCC)
Suite 401, Kyung-il Bldg
440-30 Sungnae 3-dong
Kangdong-gu, Seoul 134-847   Korea
Tel : +82 2 508 4883~5 Fax : +82 2 568 8593
e-mail info@apscc.or.kr   web: www.apscc.or.kr

Cable and Satellite Broadcasting Association of Asia (CASBAA)
802 Wilson House
19-27 Wyndham Street
Central, Hong Kong
Tel: +852 2854 9913 Fax: +852 2854 9530
Email: casbaa@casbaa.com   web: www.casbaa.com

Global VSAT Forum (GVF)
Fountain Court
2 Victoria Square
Victoria Street
St . Albans, Hertfordshire
AL1 3TF United Kingdom
Tel:+ 44 1727 884513 Fax + 44 1727 884839
Email: info@gvf.org   web: www.gvf.org

Mobile Satellite Users Association (MSUA)
1350 Beverly Rd., Suite 115-34
McLean, Virginia 22101 USA
Tel: +1 (650) 839 0376 Fax: +1 (650) 839 0375
Email: msua@msua.org web: www.msua.org

Pacific Telecommunications Council   (PTC)
2454 S. Beretania St. Suite 302
Honolulu, HI 96826-1596   USA
Tel: +1.808.941.3789 Fax: +1.808.944.4874
Email: info@ptc.org web: www.ptc.org

The Satellite Technology Guide for the 21st Century

Society of Satellite Professionals International (SSPI)
55 Broad Street, 14th floor
New York, NY 10004 USA
Tel: +1 212-809-5199   Fax: +1 212-825-0075
Email: info@sspi.org   web: www.sspi.org

Satellite Industry Association (SIA)
1730 M Street, N.W. Suite 600
Washington, D.C. 20036
Tel. +1 (202) 349-3650 Fax: (202) 349-3622
Email: info@sia.org   web: www.sia.org

Satellite Users Interference Reduction Group (SUIRG)
P.O. Box 512548
Punta Gorda, Florida 33951-2548
Phone: 1-941-575-1277
Fax: 1-941-575-7048
Email: bobames@suirg.org   web: www.suirg.org

World Teleport Association (WTA)
55 Broad Street, 14th Floor, New York, NY 10004 USA
Telephone +1 212-825-0218 Fax +1 212-825-0075
Email: wta@worldteleport.org   web: www.worldteleport.org